THE MINERAL RESOURCES OF THE SEA

Elsevier Oceanography Series

THE MINERAL RESOURCES OF THE SEA

BY

JOHN L. MERO

Consultant
Newport News Shipbuilding and Dry Dock Co.
Newport News, Va.

Formerly with the
Institute of Marine Resources
Department of Mineral Technology
University of California
Berkeley, Calif.

ELSEVIER PUBLISHING COMPANY
AMSTERDAM - LONDON - NEW YORK
1965

ELSEVIER PUBLISHING COMPANY
335 JAN VAN GALENSTRAAT, P.O. BOX 211, AMSTERDAM

AMERICAN ELSEVIER PUBLISHING COMPANY, INC.
52 VANDERBILT AVENUE, NEW YORK, N.Y. 10017

ELSEVIER PUBLISHING COMPANY LIMITED
12B, RIPPLESIDE COMMERCIAL ESTATE
RIPPLE ROAD, BARKING, ESSEX

LIBRARY OF CONGRESS CATALOG CARD NUMBER 64-18520
WITH 74 ILLUSTRATIONS AND 43 TABLES

COPYRIGHT © 1964 BY ELSEVIER PUBLISHING COMPANY
ALL RIGHTS RESERVED

FOR HOPE, ANNE AND LAWRENCE

The sea, a boundless, inexhaustible storehouse of the material stuff of civilization. (Photo by B. J. Nixon).

PREFACE

In the fall of 1957, several professors of the University of California met at the Scripps Institution of Oceanography to plan a cooperative program between two University organizations, the Department of Mineral Technology at Berkeley and the Institute of Marine Resources at San Diego, on the recovery of minerals from the sea. They were Professors P. D. Trask, H. E. Hawkes, and P. E. Witherspoon of Berkeley and Professors C. D. Wheelock and W. H. Menard of San Diego. As a first project they chose an economic analysis of mining ocean-floor manganese nodules. I was asked to be the chief investigator on this project under the faculty guidance of Professor H. E. Hawkes. Out of that study came a report on the mining and processing of manganese nodules which definitely indicated the technical and economic feasibility of using this material as a source of various metals. In the course of that study, we learned about a number of other materials on the ocean floor that might be of economic interest. Accordingly, with the help of the Institute of Marine Resources, our program was continued and broadened to include studies of all the mineral resources of the sea. Most of the original data and information of this volume was generated during that study which was conducted at the Department of Mineral Technology on the Berkeley Campus of the University and at the Scripps Institution of Oceanography on the San Diego Campus of the University. Many of the data concerning the chemical composition were developed while the author served as the D. C. Jackling Fellow for the year, 1960, with financial support coming from the Mining and Metallurgical Society of America.

The names of all the persons who helped me during those years while I was gathering data and writing preliminary reports are too numerous to mention. But certainly of greatest help were Professors G. O. S. Arrhenius, C. D. Wheelock, R. Revelle, and E. D. Goldberg of the University of California at San Diego. The idea of the compositional regions of the manganese nodules in the Pacific Ocean came as the result of a suggestion by Professor W. H. Menard. Also of great help to me were Professors L. E. Shaffer, E. H. Wisser, H. E. Hawkes, S. H. Ward, and P. D. Trask of the University of California at Berkeley who guided much of my work at the University.

Mr. W. Riedel and Professor G. O. S. Arrhenius of Scripps, Dr. B. C. Heezen and Mr. C. Fray of the Lamont Geological Observatory, and Mr. T. Stetson of the Woods Hole Oceanographic Institution kindly located and sent to the author many samples of manganese nodules for the various tests made on them. Mr. M. Silverman and Professor C. D. Wheelock of the University of California at San Diego contributed significantly in the studies concerning deep-sea dredging.

Many individuals such as Messrs. S. Calvert, D. Owen, R. Thornberg, B. C. Heezen, A. W. H. Bé, C. Shipek, W. McIlhenny, B. J. Nixon, and N. Zenkevitch, and organizations such as Leslie Salt, Kaiser Aluminum and Chemicals, Dow Chemical, Freeport Sulphur, Hughes Aircraft, General Mills, Global Marine Exploration, Ellicott Machine, Newport News Shipbuilding and Dry Dock, and the U. S. Navy Electronics Laboratory were most helpful in supplying illustrations for the book. Photographs, not otherwise credited, were taken by the author.

Mr. J. E. Flipse and Mr. E. Anderson of the Newport News Shipbuilding and Dry Dock Company kindly read parts of the manuscript and critized them as did Professor G. O. S. Arrhenius of Scripps and Mr. W. McIlhenny of Dow Chemical Co. Dr. K. O. Emery of the Woods Hole Oceanographic Institution read the entire manuscript and suggested many significant changes in the text. But probably of greatest help was Professor C. D. Wheelock who, as director of the Institute of Marine Resources, offered the author considerable moral and financial support during the course of his studies at the University concerning marine mineral resources.

To all of these persons I owe a great debt of gratitude. By including the names of some of them here I do not intend to implicate them in any way with erroneous conclusions I may have come to or mistakes I may have made.

My thanks are also due to Dr. B. C. Heezen, *The New Scientist of London*, the *Engineering and Mining Journal*, the *Journal of Metals*, John Wiley and Sons of New York, Prentice-Hall, Inc., of Englewood Cliffs, N. J., and *Scientific American* for allowing me the use of illustrations originally published by them and to the University of Chicago Press for the use of the Goode Base Maps.

Newport News, Va. JOHN L. MERO
December 12, 1963

CONTENTS

PREFACE VII

CHAPTER I. INTRODUCTION 1

CHAPTER II. MARINE BEACHES 6

Beach deposits 6
Type of minerals expected in beach deposits, 7 – Concentration of heavy minerals in beaches, 8 – Submerged beaches, 11

Beaches now being mined 12
Diamonds from offshore Southwest Africa, 12 – Beach at Nome, Alaska, 14 – The channels of southeastern Alaska, 15 – Mining of iron sands offshore of Japan, 16 – Cement rock sands, 16 – Beaches of Ceylon, 18 – Other beaches, 18

Submarine beach exploration methods 20

CHAPTER III. MINERALS FROM SEA WATER 24

Extraction of minerals from sea water 25
Extraction of common salt, 25 – Extraction of bromine from sea water, 31 – Magnesium from sea water, 34 – Magnesium compounds, 39 – Gold from sea water, 39 – Other materials extracted from sea water, 42 – Production of minerals in conjunction with the extraction of fresh water from the ocean, 43

Economics of the extraction of minerals from sea water . . 45

New technologies for mineral extraction from the sea . . . 48
Recovery of suspended matter in sea water, 50 – The concentration of elements by marine organisms, 50

CHAPTER IV. THE CONTINENTAL SHELVES 53

Surficial deposits of the continental shelf   55

Phosphorite   57
Submarine phosphorite, 59 – Origin of the phosphorite nodules, 59 – Phosphorite deposits off the coast of California, 61 – Chemical and mineralogical composition, 64 – Distribution, 67 – Economics of mining the California phosphorite, 71 – World-wide sea floor phosphorite tonnages, 73

Other minerals on the continental shelves   73
Glauconite, 73 – Barium sulphate concretions, 75 – Organic sediments, 76 – Sand and gravel, 77 – Placer deposits of drowned river valleys, 78 – Methods of exploration for drowned river valleys, 80

CHAPTER V. STRATA UNDERLYING THE SOFT SEA-FLOOR SEDIMENTS   84

Deposits under the surficial sediments of the continental shelf   89

Vein deposits.   96

Petroleum deposits   98

CHAPTER VI. THE DEEP-SEA FLOOR   103

Depth of the oceans.   106

Sediments of the ocean floor   106
Pelagic sediments, 106 – Calcareous oozes, 111 – Siliceous oozes, 115 – Animal debris, 117 – Other minerals, 120 – Red clay, 122

Manganese nodules   127
Physical forms of ocean-floor manganese–iron oxides, 127 – Physical characteristics of the nodules, 129 – Nuclei of the nodules, 135

Environmental factors of formation of manganese nodules .   139
Associated sediments, 140 – Ocean-floor currents, 142 – Effects of animals on the ocean floor, 144

Formation of manganese nodules   145

Mineralogy of manganese nodules   151
Rate of formation of the nodules, 153

Distribution and concentration of manganese nodules . . .   155
Surface concentration measurements on manganese nodules, 165 – Tonnage estimates, 174 – Rate of accumulation of manganese nodules, 175

Chemical composition of manganese nodules 178
Method of analysis, 222 – Regional variations in the composition of the manganese nodules, 225 – A-regions (high iron), 225 – B-regions (high manganese), 227 – C-regions (high nickel and copper), 228 – D-region (high cobalt), 229 – Other transition zones, 230 – Analyses of various shells within a single nodule, 230 – Closely spaced sampling of a manganese nodule deposit, 231

Amounts of various metals in the manganese nodules . . . 234

Manganese nodules in the Atlantic and Indian Oceans . . . 234
Manganese nodules in the Indian Ocean, 236

CHAPTER VII. OCEAN MINING METHODS 242

The mining of marine beaches and offshore placers 243
Wire line methods, 245 – Bucket-ladder dredges, 247 – Hydraulic dredges, 249 – Air-lift dredges, 251 – Mining minerals from the sub-seabed strata, 252

The mining of surficial sediments from the ocean floor . . 252
The deep-sea drag dredge, 253 – The deep-sea hydraulic dredge, 260 – Fluid velocity necessary to carry the nodules in the pipe line, 263 – Power required to operate the dredge, 264 – Suction heads, 265 – Production rate, 267 – Capital and production costs, 268 – Influence of weather on a mining operation, 271 – Processing of the manganese nodules, 271

CHAPTER VIII. SOME ECONOMIC AND LEGAL ASPECTS OF OCEAN MINING 273

The manganese nodules as a mineral resource 277

Advantages of deep-sea mining 280

Legal problems involved in ocean mining 280
Law pertaining to deposits of the deep sea, 284

APPENDIX I. STATION OF SAMPLE TITLE LIST 294

APPENDIX II. TABLE OF CONVERSION FACTORS 295

REFERENCES 296

INDEX 305

CHAPTER I

INTRODUCTION

The sea is generally accepted by scientists as the locale of the origin of life on earth. Without the sea, life, as we know it today, could not exist. The functions of the sea in relation to earthly life are numerous. It acts as a great thermostat and heat reservoir, leveling out the temperature extremes which would prevail over the earth without its moderating influences. It is the earth's water reservoir and supply without which the continents would be lifeless deserts. The sea provides a means for the least expensive form of transportation known to man. It is a playground for mankind; a major source of his food as well as a dumping ground for his garbage. And the sea is a major storehouse of the minerals which serve as the foundation of an industrial society.

As a source of minerals, the sea has been little exploited relative to its potential. The major reasons for this default are, I believe, a lack of knowledge concerning what is in the ocean and of the advantages of exploiting marine mineral deposits, the absence of a technology to economically exploit the deposits, and no pressing need, either economic or political, to exploit them at the present time.

In regard to mineral resources, the sea can be divided into five regions: marine beaches, sea water, the continental shelves, surficial sediments, and the hard rock beneath the surficial sea-floor sediments. A variety of minerals are presently being extracted from the first three regions of the ocean. As a result, a rather voluminous literature exists on what is being recovered and the methods used to recover them. Very little is known about the fifth region, that of the hard rock underlying the soft ocean-floor sediments. No sample of this rock has ever been obtained for chemical analysis although the thickness of the layer between the unconsolidated sediments and the mantle of the earth averages about 3 km. The emphasis of this volume, consequently, will be on the fourth region, the surficial sea-floor sediments. Such emphasis is proper for it is the fourth region that has recently been found to contain vast mineral resources of great economic promise. In addition, it is a region from which few, if any, minerals are now being taken.

For purposes of this volume, the word "mineral" will be taken to mean those elements or compounds which are normally used or marketed in an inorganic form whether or not their mode of genesis was due to organic or inorganic processes. In the open ocean, biotic processes probably dominate in separating and concentrating the various elements which enter the sea. Both plants and animals play major roles in these processes. Materials such as calcium and silicon are extracted in large quantities from sea water by plants and animals for use in forming shells and skeletons. Other elements such as copper may be concentrated by animals for use in their metabolic processes. In addition, biota such as bacteria make a living by oxidizing certain elements such as manganese or by consuming the organic parts of complexes by which such elements are held in solution. The carried elements may then be deposited and concentrated in the body of the animal or may be converted to an insoluble precipitate and released to diffuse through the sea water, slowly settling to the sea floor. After the death of the animal or plant and dissolution of the biogenous material, the residue, after diagenetic changes, can be classified as inorganic for purposes of consideration as economic mineral deposits.

The term "mineral resource" is somewhat nebulous in meaning. Economic systems, technology, political climates, governmental decisions, taxes, and a myriad of other factors are involved in the equation with which we attempt to determine whether or not a mineral deposit is a mineral resource. Sometimes, even market economics gets involved, but, seemingly, not too often any more. Mineral deposits may be economic mineral resources, or at least potential mineral resources, in one location but not in another, at one time in history but not in another, and by political decision in one country but not in another. By governmental decision and subsidy, deposits such as the Oregon nickel deposits or the Nevada manganese deposits, become mineral resources, when, under normal circumstances of free trade, they could not be considered mineral resources. We have mineral resources, which given continuing technical and economic development, will apparently last for thousands of years. Such are the phosphate reserves of western United States. A change in economic or technical conditions, however, can change the status of these reserves in a very short time. Many deposists of uranium will become uneconomic in the late 1960's when marketing contracts with the U. S. Atomic Energy Commission expire. The great, low-grade copper deposits of western United States, once considered of no value, have, through the foresight and technical and financial skills of a few men, become one of the truly great mineral resources of the world.

If minerals are being extracted from the marine environment at the present time by enterprises operating under relatively normal, free-trade conditions, additional deposits of these minerals occurring in approximately the same form as those being extracted will be considered mineral resources. Such are magnesium, salt, and bromine, which are taken from sea water, and sulphur, which is taken from deposits under the waters of the Gulf of Mexico. If it appears, that with existing technology and free-trade economics, we can mine, process, and utilize, the mineral deposits in the sea, those deposits will be classified as potential mineral resources. Most of the deposits described in this volume will fall into this latter classification.

A number of mines being operated at the present time extend out under the ocean. By and large these mines are worked with the same methods used in mines entirely within the vertical confines of the continents. Also, a large number of the minerals being mined from continental sources are marine in origin, that is, they were deposited in shallow seas and bays which at some time in geologic past engulfed the land. Practically all sedimentary mineral deposits fall into this category.

In general, discussion in this volume will be limited to those deposits which are of the present marine environment or are mined by having to work in a truly marine environment, that is, in having to deal with ocean phenomena such as waves, surf, or marine organisms.

Bodies of water, such as the Dead Sea or the Caspian Sea, which are not physically connected to the ocean will not be considered although these bodies of waters are great producers of minerals.

There is, in many cases, no sharp line between what is a marine resource and a continental resource. Petroleum deposits which lie in the continental shelf more than 2 miles from the shore are exploited by drilling directional holes from onshore drilling sites. Even though the oil deposits are marine in origin and are now located in what might be called a marine environment, it is difficult to see how these deposits can be called marine resources. Exploiting these deposits from man-made islands or drilling platforms, either resting on the bottom or floating, and thus having to deal with the sea, makes classification of the deposits as marine resources somewhat less questionable.

It probably is more proper to classify as marine mineral resources those deposits which may have been formed by continental processes above sea level, but which must be exploited by dealing with some aspect of the marine environment, such as placer deposits in drowned river valleys, rather than those deposits which may have been deposited through the action of the sea or on the sea floor, but which are now parts of the continents and can be exploited by conventional mining methods. Thus, the method of exploitation and, more specifically, the environment under which the deposit is exploited become the important factors in determining whether or not the deposit is termed marine or continental. Any vein-type deposits or salt dome deposits within the continental rocks of the continental shelf or terrace which must be exploited by offshore techniques will be considered marine resources in this volume.

One general disadvantage of the continental ore deposits of the world is their inequitable distribution. The cost of transportation after the mineral is extracted from the earth is, in many cases, the determining factor in making a mineral deposit economic to mine. Political subdivision of the earth's surface measurably complicates this nonuniform distribution of the ore deposits of the earth's crust as far as mining economics is concerned. There is, in general, sufficient mineral resources in the earth's crust to support a world population of any reasonably predicted number. The problem with ore deposits, however, is not the total quantity available in the continental rocks but their uneven distribution and mankind's propensity for indulging in political and economic systems which inhibit the free trade of these mineral commodities. Also, centers of population are rarely ever located near centers of substantial mineral production. Nationalism, which has unfortunately taken hold to an excessive degree in many of the newer nations of the world and not a few of the older ones, can cut societies off from vital mineral commodities. One of the advantages of many of the mineral deposits of the sea is that they are generally equatibly distributed throughout the oceans of the world and are available to most nations that might wish to mine them.

CHAPTER II

MARINE BEACHES

Beaches are an interesting area of the ocean from a mineral resource standpoint. Not only are they generally pleasant places in which to work but, because of the crushing, grinding, and concentrating action of the ocean surf, much of the processing of these mineral deposits has been done by nature. What mining and processing remains is, in general, simple and inexpensive. Mineral deposits that would normally be considered very low-grade, can be frequently worked at a profit if in beach deposits. It is because of the action of the ocean surf that certain valuable minerals are concentrated in marine beaches in such a manner that allows profitable extraction; thus, these beaches can be considered true marine mineral resources, especially when surf and sea water must be dealt with in the mining of them.

BEACH DEPOSITS

Normally, the bulk of the material forming the beach is washed down to the sea coast by rivers and streams (KUENEN, 1950). This material is derived from continental rocks by weathering processes within the drainage basin of the river or stream. In certain areas, the sea floor adjacent the beach may be a major source of sand for the beach. Some additional material is derived from the erosion of rocks which crop out along the coast (JOHNSON, 1919). In certain dry areas, such as the west coast of North Africa, wind blown material may form a considerable portion of the beach material. After reaching the sea coast, the sand grains may be moved longshore or seaward by wave and surf action.

Material moved along a coast by a predominately unidirectional impingement of waves on the coast may ultimately be swept into a canyon, the head of which may be close to shore, and out to sea along the bottom of the canyon. The steady flow of sand over the edge of offshore cliffs has been noted in several places at the southern end of Baja California indicating a substantial longshore movement of beach sands in the area.

In various areas, seasonal cycles prevail in which a beach may lose much of its sand in one season and regain it in another. Along the coast of California beach sands are normally moved seaward or in a southerly direction during winter months and shoreward or in a northerly direction during spring and summer months. This constant motion of beach sands, of course, allows many opportunities for the separation of sand grains of different densities and the concentration of heavy minerals generally at the bottom of the sand layer.

Type of minerals expected in beach deposits

The same type of minerals can be expected to be found in beach deposits that are found in placer deposits on shore. That is, those minerals that are resistant to chemical and mechanical weathering processes. Easily soluble minerals such as chalcanthite cannot be expected to be found in beach deposits nor can minerals that have little resistance to mechanical erosion such as gypsum, unless, like calcium carbonate in many tropical areas, there is little other material exposed to the grinding action of the surf and available to form beach sand. On the larger volcanic islands, black sand beaches are sometimes formed by the erosion of black volcanic rocks which reach the sea coast in lava flows.

Although micas, feldspars and other silicates, and quartz form the bulk of the material in most beaches, considerable quantities of minerals such as columbite, magnetite, ilmenite, and zircon are commonly found in beaches. Less commonly found in mineable concentrations are gold, diamonds, cassiterite, scheelite, wolframite, monazite, and platinum. All of these minerals would be classified as heavy minerals (specific gravity greater than that of bromoform, 2.85), and all are generally resistant to chemical weathering.

A study of the geology and mineralogy of the rocks in the drainage areas of rivers providing sediments for a particular beach will normally indicate what valuable minerals can be expected to be found in the beach. It is logical to assume that the diamonds found in the beaches of the west coast of southern Africa were carried there by the Orange River which drains many of the diamondiferous areas of the continent. Additional evidence of an inland source for the diamonds is indicated by the placers which occur along the Orange River and by the decreasing size gradation of the diamonds north and south of the mouth of the Orange River. There is, however, considerable speculation concerning the origin of these diamonds with a number of different theories including one claiming the existence of volcanic pipes, at least at some time in the past if not evident now, offshore or in the immediate vicinity of the beaches in this area.

The apparent source of the gold in the beaches at Nome, Alaska, is the schists of the Anvil Creek drainage basin. Erosion processes free the gold from these rocks and carry it to the coast. It is dispersed by wave action along this coast and subsequently concentrated in certain horizons within the beaches. A number of beaches are found in this area at various elevations above sea level. The level of the sea was substantially lowered during the Ice Ages as the result of water piling up in the continental glaciers (KUENEN, 1950; SHEPARD, 1963). During the intermediate warm periods sea level probably experienced risings in excess of the present elevation of the sea surface. At each level of the sea which was maintained for any length of time, a new beach could be expected to be formed. Such is probably the case at Nome. At Nome, the gold values were greatest in the highest and farthest inland beach as would be expected for this beach would undergo the least amount of reworking on subsequent changes of sea level.

Concentration of heavy minerals in beaches

The jigging action of the ocean waves and surf tends to concentrate the heavy minerals in certain zones of the beach. As indicated in Fig. 1, these areas are generally in the rear of the backshore or along the contact between the unconsolidated beach sand and the consolidated basal rocks. A breaking wave as shown in Fig. 2 can pick up all available foreshore materials indiscriminately and throw them up on the beach while the backwash, not having the carrying capacity of a breaking wave, will preferentially wash the light minerals seaward. A concentrate of heavy minerals is thus generated in the rear of the backshore area (RAO, 1957). Changing sea levels, tides, and irregular heights of the surf, however, work to modify this particular concentrating action of the surf. Stringers of heavy minerals often can be found throughout the beach. Because of the presence of a lee area, near a large rock or other obstruction, the heavy minerals will tend to concentrate near this object.

Concentrations of the heavy minerals in the bottom layers of the sand of a beach are also common, especially if the base of the beach consists of a rock with many fissures and crevices into which these minerals can be deposited.

Wind also plays a role in forming concentrations of heavy minerals in beaches. In many localities, wind from a particular direction, generally toward shore, will predominate. The lighter materials are blown shoreward into dunes. Heavy minerals are left behind on the beach.

Longshore currents can work to concentrate heavy minerals in the offshore zones by winnowing out the lighter materials and transporting them away from the deposit area. Tidal currents may also function in a similar manner, however, the expected result would be the concentration of the heavy minerals at the base of the unconsolidated sediment layer over which the currents are operable.

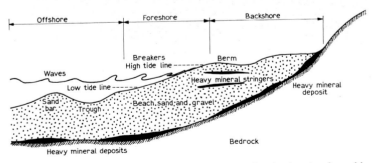

Fig. 1. A generalized, cross-sectional view of a marine beach, showing favorable locations for the accumulation of heavy minerals.

Fig. 2. The carrying capacity of a breaking wave is considerable. As much as several percent of the volume of the wave may be suspended solids being carried shoreward by the breaking wave. After the wave has spent its energy on the beach, the backwash preferentially sweeps the lighter material back to the sea, leaving a concentration of heavy minerals on the beach. (Photo by B. J. Nixon).

Seaward of the surf zone, the water wave motions provide a jigging action which tends to concentrate the heavy minerals at the base of the loose sand and gravels. As the crest of a wave passes over the foreshore section of a beach, a compressional wave is impressed on the bottom sediments. As the trough of the wave passes over the sediments, they experience a rarefaction in pressure. These alternating forces disturb the sediments and impart to them a jigging motion which allows the heavy minerals to migrate to the bottom of the unconsolidated layer where they are concentrated.

Submerged beaches

In addition to the beaches which we now find high above present sea level and those presently at sea level, we can expect to find a number of submerged beaches in the offshore area. The submerged beaches formed during the Ice Ages when sea level was much below its present level. As we would expect that the valuable, heavy minerals would be concentrated in the bottom layers of the sediments of these beaches, these layers should have been left relatively intact when the level of sea rose. Assuming that the composition of the rocks in the drainage area of the river supplying this section of the coast was the same in the past as at present, we could expect to find the same kinds of minerals in these offshore beaches. This assumption, of course, holds only if the drainage patterns have not changed during the time span involved and the chemical character of the rocks has not changed as the weathering and erosional processes worked at successively lower layers.

A study of the present beaches in any area, thus, should provide valuable information concerning what can be expected to be found offshore. The fact that beaches occur offshore can be determined by modern methods of seismic surveying (OFFICER, 1959; BECKMANN et al., 1962). These seismic methods will also yield information concerning geomorphic form of submerged beaches, depth of burial under modern surficial sediments, as well as thickness of the sediments of the beach (Fig. 4), and, of course, depth of the overlying water. Determination of geomorphic form is important for knowing it helps to guide placement of subsequently drilled exploration holes. In general, with submerged beaches assuming no catastrophic change in sea level during formation of the beach, we can expect to find the valuable mineral concentrations at the contact between the beach sands and the underlying hard rock.

BEACHES NOW BEING MINED

Table I is a selected list of world beaches that contain mineral deposits of economic interest. Under normal circumstances submerged beaches containing deposits of similar minerals should be found in the offshore areas. In general it can be expected that the offshore beaches will be greater in size than the present sea-level beaches.

Diamonds from offshore Southwest Africa

Diamonds have been mined for decades from the raised beach sands along the Atlantic coast of Southwest Africa. Geologists employed by

TABLE I
SELECTED LIST OF MARINE BEACH DEPOSITS

Location	Minerals present[1]	References
Natashquan, Quebec	M, I	DULIEUX (1912)
North Carolina	HM	MCKELVEY and BALSLEY (1948)
Florida	HM	MARTENS (1928), PHELPS (1940)
Costa Rica	HM	CRUICKSHANK (1962)
Brazil	HM, Mz	GILLSON (1951)
Argentina	HM	
Southern Chile	HM, Au	CRUICKSHANK (1962)
Guatemala	HM	BOOS (1940)
Baja California	P	
Redondo, California	HM, M	LYDON (1957)
Monterey, California	QS	
Southwest Oregon	C, Au, Pt	GRIGGS (1945), TWENHOFEL (1943)
West Kodiak, Alaska	Au, M	MADDREN (1919)
Nome, Alaska	Au, M	GIBSON (1911)
Egypt	HM, Mz	HIGAZY and NAGUIB (1958)
Senegal	HM	PARTRIDGE (1938)
Southwest Africa	HM, D	COLLINS and KEEBLE (1962)
South Africa	HM, I	COETZEE (1957)
Tranvancore, India	HM, Mz	TIPPER (1914)
Northeast Ceylon	HM	CRUICKSHANK (1962)
Western Formosa	HM	CHEN (1953)
Queensland, Australia	HM	CARLSON (1944), BEASLEY (1948)
New South Wales, Australia	HM, C	BLASKETT and DUNKIN (1948)

[1] HM = Heavy Minerals (Magnetite, ilmenite, zircon, rutile, monazite); QS = Quartz Sands; P = Phosphorite; M = Magnetite; I = Ilmenite; D = Diamonds; Au = Gold; Pt = Platinum; C = Chromite; Mz = Monazite.

the diamond mining companies have always speculated on the existence of diamonds in the sea-floor gravels offshore of this area. Inundation of below-sea-level onshore deposits with sea water, which rendered these deposits unworkable, made it seem unlikely that deposits offshore would ever be exploited, even if they did exist.

Eventually a group with no little initiative became interested in this area and started to prospect the offshore sands and gravels. Diamonds were found in a mixed alluvium of sand, gravel and boulders in water depths ranging up to 100 ft. One 4.5-ton sample, taken in 1961, is said to have yielded nine carats of diamonds with a value of U.S.$ 450. Concessions on these deposits were obtained by Mr. S. V. Collins of The Marine Diamond Corp. These concessions stretch from the mouth of the Orange River to Diaz Point near Lüderitz in Southwest Africa. The concession area is defined by the South Africa Government as extending from the low water line to a point 3 miles at sea.

Results of the initial prospecting venture, in 1961, were so promising that work on the design of a recovery method was immediately started. Within a year, a dredge was in operation mining diamonds from the gravels in this area. The diamondiferous gravels are removed from the sea floor by means of an air-lift hydraulic dredge. The system consists essentially of a large-diameter pipe into which air is pumped. The air displaces water within the pipe line resulting in a lessening of the overall density of the fluid column within the pipe line. As water at the bottom rushes into the pipe line it carries along with it the diamondiferous gravels. The bottom sediments are stirred up with water jets to increase the efficiency of the operation.

On the mining barge, the oversize and water are separated from the material dredged with the undersize undergoing further concentration. A final concentrating step using a heavy-media-separation process preceeds collection of the diamonds on a grease belt. An additional step is required before running the heavy-media concentrate over the grease belt, that of conditioning the concentrate with caustic soda and fish oil. The diamonds stick to the grease belt, while most of the other material does not. The recovery of diamonds from the dredged material is claimed to be quite good, however, some difficulties have been experienced in adequately cleaning the bedrock where the highest values of diamonds are apparently concentrated. This inability to clean the bedrock is a commonly experienced disadvantage in using a hydraulic dredge to mine placer deposits.

The offshore diamond-mining plant is completely self-contained, carrying supplies for a two-month period and operating a sea-water-conversion unit. The operation requires a complement of 60 men. In full production, the plant is capable of treating about 18,000 tons of diamondiferous material per month (COLLINS and KEEBLE, 1962). Production rates of 700 carats per day of gem quality diamonds were claimed for this operation in late 1962.

Beach at Nome, Alaska

Beaches near Nome, Alaska, have been mined for over 60 years for gold (GIBSON, 1911). Several old beaches are found at different elevations above sea level in this area. They range in age from Pliocene through the Pleistocene (HOPKINS et al., 1960). The sediments in these emerged beaches consist of iron-stained quartz, magnetite, garnet sand, and of gravel composed mainly of well-rounded quartz pebbles locally cemented by calcium carbonate. The present sea-level beach at Nome is about 200 ft. wide and slopes gently out to sea.

Early in 1961, the Shell Oil Company applied for, and received from the State of Alaska, a permit allowing them to explore the sea floor offshore of Nome. This permit covers an area of 8 square miles. At the seaward end of the Shell permit area, the water is about 60 ft. deep. During the summer of 1962, a seismic and magnetic survey was made of the area. The sparker records showed a much steeper sloping bedrock than was presumed by projecting the bedrock profile from shore. As a result, the offshore beach thickness was much greater than anticipated, approaching 300 ft. in some areas. Anomalies were found in the magnetic records which indicated the presence of magnetite accumulations in these offshore sediments. Magnetite is found with the gold in this area onshore; thus, there seems to be good possibilities of finding substantial deposits of gold in the offshore area.

Because of the shortness of the working season at Nome, sampling of this deposit was postponed for a future date. The plan presently is to drill these deposits either from the ice cover in the winter or from a floating drilling platform in the summer. Because of the climate in this area and the depth of the overburden of these deposits, sampling operations will be challenging.

In addition to the Shell permit, 27 other permits were granted to individuals for prospecting rights in Norton Sound off Nome. Such prospecting permits can be obtained merely by requesting them and paying a fee of U.S.$ 20. These additional permits were probably taken on a gamble that Shell would find mineable deposits of gold in the offshore area.

The channels of southeastern Alaska

A number of permits have been granted to explore and dredge for gold in the channels of southeastern Alaska. Williams Hydraulics Incorporated of Oakland, California, has plans to explore for gold in Stephens Passage south of Grand Island. The permit area again covers 8 square miles with water depths of 520–900 ft. Gold was noted in the deposits of this passage during sediment sampling operations before the laying of telegraph cables to Alaska. Generally the gold in these deposits is so fine, that any processing method short of dissolution fails to recover any significant amount. Mr. J. C. Williams, president of the company that holds this permit, has developed a 5-ton capacity clamshell bucket which uses the pressure of the water at the bottom of the channel to actuate its closing mechanism. No significant amount of gold has ever been recovered from these channel deposits, however, permits to explore the deposits, which are convertible into mining leases if mineable deposits are located, only cost U.S.$ 20 and no more pleasant place to explore for gold, at least in the summer when the salmon are running, can be found in this world. Many other places in Alaska show greater promise for containing economically mineable deposits, generally where gold bearing streams empty into tide water basins. Tidal currents in many areas of Alaska, however, are fierce to say nothing of the climate in the northern areas. Deposits would have to be high-grade to repay the costs of mining under such circumstances. But if they are found, they probably will be mined for such is the lure of gold.

Mining of iron sands offshore of Japan

Magnetite is a common constituent of beach placer deposits. The magnetite content of most beaches, however, is too small to warrant economic mining. Off the southern tip of Kyushu Island, Japan, however, beach deposits of magnetite have been mined for many years. Recently, mining has moved to the offshore area. In Ariake Bay, the Yawata Iron and Steel Company has blocked out nearly 40 million tons of refinable iron ore reserves in a 31 square mile area of the offshore beach sands. The Ariake Bay deposits assay 56% iron, 12% titanium oxides, and 0.26% phosphorus (ANONYMOUS, 1962). Exploration of these deposits is accomplished with magnetometers and by coring.

Several systems of mining these underwater sands have been tried including hydraulic, bucket line, and grab-bucket methods. Of the methods tried, the grab-bucket methods proved to be the most effective. High water currents in Ariake Bay made it difficult to control the vessel with concomitant deleterious effects on hydraulic or bucket-line methods which depend on constant contact with the bottom to be effective. In still water, the hydraulic or bucket-line methods might be expected to be the most efficient methods.

A pilot grab-bucket dredge with a 10.5-cubic-yard bucket has been designed, built, and put into operation in this deposit. Its production rate is reported to be about 30,000 tons per month. The cost of recovering iron ore from these sands is reported to be about U.S.$ 5 per ton of concentrate which cost covers dredging and concentration aboard the dredge. About 5% of the material dredged is recoverable iron ore.

Cement rock sands

In a number of locations shell sands, formed by the mechanical erosion of calcareous shells of animals, are mined to be used in the production of Portland cement or of lime. Oyster shells are used as the basic raw material for the manufacture of cement by the Ideal Cement Company, operating in the San Francisco Bay area. Other shell deposits are mined offshore Louisiana and Washington. Dow Chemical dredges oyster shells from the Gulf of Mexico, for the manufacture of lime to be used in the precipitation of magnesium from sea water. Probably the most interesting of the offshore shell mining operations, however, is that offshore western Iceland.

In 1949, a systematic survey of the island of Iceland for limestone to be used in a domestic cement industry located a deposit of shell sands about 10 miles offshore in Faxa Bay. As indicated in Fig. 3, Faxa Bay is located in the southwest part of the island. The deposit was blocked out by the use of a grab sampler. A piston corer was used to assess the thickness of the deposit which was found to range from 3–13 ft. The sand, as dredged, is a mixture of shell particles and basaltic tuff. It assays about 80% calcium carbonate (CRUICKSHANK, 1962). The sea bottom to the west of the deposit area is relatively shallow and rocky and has a rich growth of molluscs. During the winter months, the waves from the Atlantic remove, crush, and transport the shells of these molluscs toward the east into Faxa Bay. The sea floor deepens in Faxa Bay so the sediments are transported downhill. The maximum depth in the deposit is about 130 ft. of water. Near shore, the tidal currents turn the flow of shells southward and presumably back out to sea. More material is carried into the area each year than is removed in the mining operation so the deposit is self-replenishing.

The deposit is mined by means of a hydraulic suction dredge. As the deposit material is unconsolidated and free-flowing, no cutter

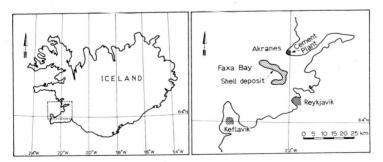

Fig. 3. Map of Iceland, showing the calcareous shell deposit on the floor of Faxa Bay. This deposit is mined to provide a cement plant near Akranes with the primary raw material for the manufacture of Portland cement.

head is necessary in the dredging operation. The suction line is 24 inches in diameter and 160 ft. in length. Hydraulic controls operate to keep the suction head on the ocean floor through a 6 ft. rise or fall of the vessel. On the average 8,000 tons of water are pumped per hour carrying 3–5% solids. The dredged solids are separated from the water and stored in the hold of the dredging vessel. The vessel can carry 1,000 tons of sand. When full, the vessel moves to port and pumps the cargo ashore.

Beaches of Ceylon

In northeastern Ceylon, beaches have been worked for over 60 years. The major constituents of the beach sands in this area are ilmenite (75–80%), rutile (6–10%), zircon (6–7%), and magnetite (2–3%). The source of the sand is the metamorphic rock complex inland from the beach areas. Normal erosional processes free the minerals from these rocks and transport them to the sea coast. The mining of these beaches is seasonal and storms generally replenish the valuable mineral content of mined areas with minerals from the offshore area (CRUICKSHANK, 1962). Thus, the areas offshore of these beaches are effectively mined even though the physical mining operation is confined to the fore- and backshores of the beaches.

Replenishment by wave action after they have been mined is a common feature of marine beach deposits. In fact, continuing replenishment of the mined material, as we shall see, is a common feature of many mineral deposits of the sea.

Other beaches

The beaches of western North America contain magnetite, gold, platinum, and monazite almost without exception, although few in mineable concentrations. Some of these beaches have already been mined. In 1927 and 1928, Redondo Beach, California, was mined of about 15,000 tons of ilmenite concentrates (LYDON, 1957). The concentration of ilmenite in the beach was about 7%. Beaches in southwestern Oregon are presently being mined for chromite and other heavy minerals including gold and platinum (TRUMBULL, 1958). The beach at Monterey, California, is mined for silica sand and the beach at El Segundo, California, is mined for sand used in sand blasting and foundry molds. Beach gravels between Carlsbad and Encinitas, California, were mined for grinding stones and as filter aids until 1948 (TROXEL, 1957).

The beaches of southwestern India, Queensland and New South Wales, Australia, and southern Brazil, have been worked for many years for their ilmenite, rutile, monazite, and zircon (BEASLEY, 1948; BLASKETT and DUNKIN, 1948; GILLSON, 1951; TIPPER, 1914). Recently, deposits of titanium minerals in beach sands along the eastern Florida coast have been developed and are being worked. The Taranake deposits on the southwest shore of the North Island of New Zealand contains millions of tons of heavy minerals which assay about 9% TiO_2 (AUBEL, 1920).

It is generally safe to assume that similar deposits of minerals exist in the offshore beaches as in those beaches on shore. The technology of exploring offshore deposits is already developed in the systems used to sample offshore sediments for oil. A mining technology is presently being developed for the exploitation of such deposits. A major difficulty in mining beach deposits, however, is that the pay-streak is generally at the contact between the unconsolidated sediments and the bedrock. If the bedrock surface is fissured or potholed, the valuable materials tend to collect in the fissures and holes. To efficiently mine these deposits, the mining device should be able to cut into the bedrock a short distance. If the bedrock is slightly weathered or is composed of relatively soft rocks and not submerged more than about 100 ft., and if the area is relatively protected from ocean waves and swells, the ladder-bucket excavator can efficiently mine the deposit. An adequate job of cleaning the bedrock is not generally possible with the other methods of dredging these sediments. A major improvement in the methods of mining beach deposits could be realized if some method of undercutting the overburden and taking only the contact zone of the deposit could be devised. Preliminary designs of such machines have been developed, but, as yet, none of them have been built or tested on a commercial scale.

As methods of mining the offshore beach deposits are improving, we can expect to see an increase in mining activity in these deposits. In Japan, methods of using titaniferous iron ores in the steel-making process have been developed which development obviates a major objection of utilizing the rather substantial deposits of magnetite in marine beaches. Recently, the Ontario Research Foundation of Toronto, Canada, announced the development of a process in which natural gas is used to smelt magnetite sands (ANONYMOUS, 1963b). It is indicated that the process, when fully developed, would require about 35,000 cubic ft. of gas per ton of iron produced. Advantages of the process are: fine-grained ores can be fed to the furnace with no need for pelletizing; capital costs are relatively low; and, there is no dependence on a supply of coking coal. A particularly important advantage of this process is that it is economically attractive even when used in producing quantities of pig iron as small as 25 tons per day. It is, thus, attractive for use in underdeveloped countries that have a supply of natural gas.

SUBMARINE BEACH EXPLORATION METHODS

The exploration of submerged beaches can probably best be accomplished with methods developed for oil exploration. Primary, of course, is a study of the general geology of the area with extrapolation to the offshore area, when possible, of the known geology of the adjacent onshore areas. Initial surveys of the beach itself are probably best done with the various sonic devices now used to record bottom profiles and to delineate the bedding planes and dissimilar rock contacts to a depth of 1,500 ft. into the sediments. Fig. 4 shows a graph obtained by the use of such a device and an interpretation of the actual echogram. The method is similar to a reflection seismic survey and can utilize a variety of sonic sources to produce a low-frequency sound pulse which penetrates the water and sediment layers. An echo is generated by the primary sound pulse at each marked change in density of the sediment. The depth below sea level of the contact between the unconsolidated sediments and the bedrock or between rocks of different densities can thus be determined. In many cases sediment types also can be distinguished from these

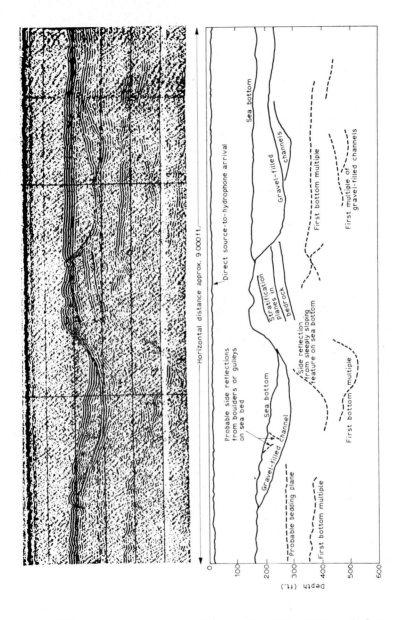

Fig. 4. Echogram of a sonic survey, illustrating the ability of this method to delineate subsurface sediment structures. The lower part of the illustration is an interpretation of the echogram. (After BECKMANN, 1960. Printed by permission of *The New Scientist*).

records. The graphs are especially useful in locating sediment traps, drowned river valleys, and, in the case of submerged beaches, geomorphic forms.

Parallel traverses of the deposit area are made with these devices carried either by vessels or by helicopters and the various bedding planes and geomorphic contours of adjacent traverses can be correlated to obtain a three-dimensional representation of the deposit components (BECKMANN, 1960; BECKMANN et al., 1962; MARIES and BECKMANN, 1961).

Magnetometer surveys of submerged beaches are also important as most economically significant concentrations of beach placer minerals contain substantial percentages of magnetite. The magnetic anomalies generated by the magnetite accumulations are generally very apparent on the graph. Magnetic surveys are especially important in the case of prospecting for gold. Normally they can be run in conjunction with the sonic survey.

Some preliminary surficial sampling of the beach should be done to pick up mineral dispersion patterns if such occur in the area, to establish the surficial rock compositions, and to gain some information concerning the physical characteristics of the material in the beach. Following the presentation and interpretation of the data of the sonic, magnetic, and geologic surveys is physical sampling of the deposit at depth. Drop corers and clamshell buckets can be used effectively in sampling the upper few feet of the surficial sediments. Scuba equipment and divers can also be used for this operation. The bedrock can be sampled with a jetting corer if it is not too deeply submerged. The jet corer consists of a length of pipe which is lowered from a vessel. High velocity water is pumped through the pipe and the jetting action of this water issuing from the lower end of the pipe very effectively cuts a hole in the unconsolidated overburden sediments. Once at bedrock, the pipe is rammed into the rock with sufficient force to obtain a plug several inches in length.

Coring with a drill is probably the only effective way of adequately sampling a cross-section of a beach deposit. Obtaining complete or representative cores in the generally unconsolidated beach deposits, however, is difficult and close supervision of the drilling operation must be maintained to avoid contamination of the sample by side-wall sluffing. Any hole drilled in beach material, if greater than a few tens of feet, normally must be cased. The technique of drilling holes in the ocean floor is well developed and may be accomplished with bottom supported rigs in water depths less than about 100 ft. or by floating drilling vessels in any depth of water overlying a beach deposit sufficient to float the vessel.

CHAPTER III

MINERALS FROM SEA WATER

If whoever named the planet were aware of the true nature of its surface, they would probably have given it some name denoting water rather than earth. Almost 71% of the earth's surface is water. Covering an area of 139 million square miles at a mean depth of 2.46 miles, the sea holds about 330 million cubic miles of water (SVERDRUP et al., 1942). Sea water contains an average of 3.5% of various elements in solution; thus, each cubic mile of sea water, weighing some 4.7 billion tons, holds about 166 million tons of solids. If all the salt in the ocean could be recovered and stored on the continents, it would cover them to a depth of about 500 ft. The oceans are a storehouse of some $5 \cdot 10^{16}$ tons of mineral matter.

Given sufficiently sophisticated analytical procedures, probably every known, naturally occurring, element could be found in sea water. The concentration of some 60 elements in sea water has been measured, and Table II lists these elements, their concentration, the amount of the elements in one cubic mile of sea water and the total amount in the oceans of the world. Other elements such as ytterbium, beryllium, zirconium, platinum, etc., can be inferred to be part of sea water because of their appearance in authigenic minerals of the sea floor or in marine organisms.

The concentrations listed in Table II suffer from a number of deficiencies. Many of the values given were obtained from a single set of analyses of a sample of surface water. The surface waters are generally much less characteristic of the hydrosphere than are samples taken from a depth of 5,000 ft. or so. Biological activity at the ocean surface can work significant changes in the concentrations of various elements from place to place and from time to time (GOLDBERG, 1963a). It is never safe to assume that sea water is an absolutely homogeneous medium where the concentration of the less abundant elements is concerned.

Two elements, sodium and chlorine, constitute 85.2% of the dissolved solids in sea water while the nine most abundant elements constitute over 99% of the total dissolved solids in sea water. These nine elements are found in remarkably constant proportion to one another (H. W. HARVEY, 1960). The chloride ion constitutes 54.8% of the total salts, the sodium ion 30.4%, the sulphate ion 7.5%, the magnesium ion 3.7%, the calcium ion 1.2%, the potassium ion 1.1%, the carbonate ion 0.3%, the bromide ion 0.2%, and the borate ion 0.07%.

EXTRACTION OF MINERALS FROM SEA WATER

Of the 60 or so elements known to be dissolved in sea water, only four have been commercially extracted in any quantity. They are sodium and chlorine in the form of common salt, magnesium and some of its compounds, and bromine. Several calcium and potassium compounds are produced as by-products in salt- or magnesium-extraction processes or by processing seaweeds, which extract these elements from sea water and concentrate them in their cells, however, these elements are not commercially extracted directly from sea water itself on any notable scale. Attempts have been made to extract other minerals from sea water, but they have not been succesful as commercial enterprises. In addition to many patents being granted on processes for the extraction from sea water of common salt, magnesium, magnesium compounds, and bromine, patents have been granted covering the extraction of iodine, potassium, calcium sulphate, gold, and silver (BAUDIN, 1916; CERNIK, 1926; NICCALI, 1925; S. O. PETTERSON, 1928; VIENNE, 1949).

Extraction of common salt

Salt was apparently first systematically extracted from sea water by the Chinese some time prior to 2200 B.C. Since that time many other societies have depended on sea water as a source of salt (ARMSTRONG and MIALL, 1946). Today, solar salt is an important part of the total

TABLE II

CONCENTRATION AND AMOUNTS OF 60 OF THE ELEMENTS IN SEA WATER

Element	Concentration (mg/l)[1]	Amount of element in sea water $(tons/mile^3)$	Total amount in the oceans $(tons)$
Chlorine	19,000.0	$89.5 \cdot 10^6$	$29.3 \cdot 10^{15}$
Sodium	10,500.0	$49.5 \cdot 10^6$	$16.3 \cdot 10^{15}$
Magnesium	1,350.0	$6.4 \cdot 10^6$	$2.1 \cdot 10^{15}$
Sulphur	885.0	$4.2 \cdot 10^6$	$1.4 \cdot 10^{15}$
Calcium	400.0	$1.9 \cdot 10^6$	$0.6 \cdot 10^{15}$
Potassium	380.0	$1.8 \cdot 10^6$	$0.6 \cdot 10^{15}$
Bromine	65.0	306,000	$0.1 \cdot 10^{15}$
Carbon	28.0	132,000	$0.04 \cdot 10^{15}$
Strontium	8.0	38,000	$12,000 \cdot 10^9$
Boron	4.6	23,000	$7,100 \cdot 10^9$
Silicon	3.0	14,000	$4,700 \cdot 10^9$
Fluorine	1.3	6,100	$2,000 \cdot 10^9$
Argon	0.6	2,800	$930 \cdot 10^9$
Nitrogen	0.5	2,400	$780 \cdot 10^9$
Lithium	0.17	800	$260 \cdot 10^9$
Rubidium	0.12	570	$190 \cdot 10^9$
Phosphorus	0.07	330	$110 \cdot 10^9$
Iodine	0.06	280	$93 \cdot 10^9$
Barium	0.03	140	$47 \cdot 10^9$
Indium	0.02	94	$31 \cdot 10^9$
Zinc	0.01	47	$16 \cdot 10^9$
Iron	0.01	47	$16 \cdot 10^9$
Aluminum	0.01	47	$16 \cdot 10^9$
Molybdenum	0.01	47	$16 \cdot 10^9$
Selenium	0.004	19	$6 \cdot 10^9$
Tin	0.003	14	$5 \cdot 10^9$
Copper	0.003	14	$5 \cdot 10^9$
Arsenic	0.003	14	$5 \cdot 10^9$
Uranium	0.003	14	$5 \cdot 10^9$
Nickel	0.002	9	$3 \cdot 10^9$
Vanadium	0.002	9	$3 \cdot 10^9$
Manganese	0.002	9	$3 \cdot 10^9$
Titanium	0.001	5	$1.5 \cdot 10^9$
Antimony	0.0005	2	$0.8 \cdot 10^9$
Cobalt	0.0005	2	$0.8 \cdot 10^9$
Caesium	0.0005	2	$0.8 \cdot 10^9$
Cerium	0.0004	2	$0.6 \cdot 10^9$
Yttrium	0.0003	1	$5 \cdot 10^8$

[1] After GOLDBERG (1963a).

TABLE II (continued)

Element	Concentration (mg/l)[1]	Amount of element in sea water $(tons/mile^3)$	Total amount in the oceans $(tons)$
Silver	0.0003	1	$5 \cdot 10^8$
Lanthanum	0.0003	1	$5 \cdot 10^8$
Krypton	0.0003	1	$5 \cdot 10^8$
Neon	0.0001	0.5	$150 \cdot 10^6$
Cadmium	0.0001	0.5	$150 \cdot 10^6$
Tungsten	0.0001	0.5	$150 \cdot 10^6$
Xenon	0.0001	0.5	$150 \cdot 10^6$
Germanium	0.00007	0.3	$110 \cdot 10^6$
Chromiun	0.00005	0.2	$78 \cdot 10^6$
Thorium	0.00005	0.2	$78 \cdot 10^6$
Scandium	0.00004	0.2	$62 \cdot 10^6$
Lead	0.00003	0.1	$46 \cdot 10^6$
Mercury	0.00003	0.1	$46 \cdot 10^6$
Gallium	0.00003	0.1	$46 \cdot 10^6$
Bismuth	0.00002	0.1	$31 \cdot 10^6$
Niobium	0.00001	0.05	$15 \cdot 10^6$
Thallium	0.00001	0.05	$15 \cdot 10^6$
Helium	0.000005	0.03	$8 \cdot 10^6$
Gold	0.000004	0.02	$6 \cdot 10^6$
Protactinium	$2 \cdot 10^{-9}$	$1 \cdot 10^{-5}$	3,000
Radium	$1 \cdot 10^{-10}$	$5 \cdot 10^{-7}$	150
Radon	$0.6 \cdot 10^{-15}$	$3 \cdot 10^{-12}$	$1 \cdot 10^{-3}$

[1] After GOLDBERG (1963a).

salt consumed in China, India, Japan, Turkey, the Philippine Islands, and other countries. About 6 million tons of solar salt are produced annually throughout the world. Normally, a hot, dry climate with dry winds is required for the production of salt by evaporation of sea water, and in almost all populated areas where such conditions exist, with land available and the sea near, solar salt is produced. In addition to nearness to the sea and a hot climate, the following qualifications should be met in an area where solar salt is to be produced: the soil should be almost impermeable; large areas of flat ground at or below sea level should be available; little rainfall during the evaporating months; no possibility of dilution from fresh water streams; and, because of salt's low selling price, inexpensive transportation or markets nearby.

About 5% of the United States comsumption of salt is produced by solar evaporation principally in the San Francisco Bay area where production of solar salt began in 1852. Fig. 5 shows the evaporation ponds at the south end of San Francisco Bay. These ponds, operated by the Leslie Salt Company, cover an area of about 80 square miles and produce about 1.2 million tons of salt annually. Evaporation ponds are also in operation at the heads of Newport Bay and San Diego Bay in Southern California where the annual production is about 100,000 tons (EMERY, 1960).

At the San Francisco Bay operation, sea water is taken into evaporation ponds at high tide through sluice gates in dikes which surround

Fig. 5. Solar salt evaporation ponds at the south end of San Francisco Bay, California. In the center of the photograph is shown a stockpile of recovered salt. (Photo courtesy of the Leslie Salt Co., San Francisco, California).

the pond areas. The sea water is held in these ponds until sufficient water has evaporated to start the precipitation of its salts. Calcium sulphate is the first salt to crystallize from solution. As it settles to the bottom, the remaining brine is slowly moved into the crystallizing ponds and further concentrated by evaporation until sodium chloride begans to precipitate. The brine is evaporated to a specific gravity of about 1.28 at which point magnesium salts will start to precipitate. The brine is now called bitterns and it is drawn off the crystallizing pond and further processed elsewhere for the recovery of various magnesium compounds, bromine, and other salts. As the bitterns are removed, fresh brine is fed into the crystallizing pond and the cycle is

Fig. 6. Mechanical scrapers are used to harvest the bed of crystallized salt. At the time of harvest, the salt bed is generally 4–6 inches in thickness. (Photo courtesy of the Leslie Salt Company, San Francisco, California).

repeated. By August 1, there is a layer of sodium chloride in the bottom of the pond 4–6 inches thick. It is harvested with mechanical scrapers and loaders, as shown in Fig. 6, washed with salt water to remove impurities and dumped into storage piles. The storage piles are shown in Fig. 7. No further refining of the salt is necessary for most industrial applications, however, for human consumption, the salt is further refined. The refined product is over 99.9% NaCl. In the United States, solar salt varies in price from about U.S.$ 10 per ton for crude salt at the mine to about U.S.$ 150 per ton for packaged, refined salt at the consumer's dinner table.

The process for producing salt from sea water by solar evaporation is much the same the world over, however, the amount of cheap labor available in certain countries allows modifications to take advantage of this labor.

In some countries such as Sweden and Russia, salt is obtained from

Fig. 7. Stockpile of harvested solar salt at the Leslie Salt Company's San Francisco Bay operation. For most applications the salt requires no further purification after it is harvested and washed. (Photo courtesy of the Leslie Salt Company, San Francisco, California).

sea water by freezing the water. The ice that forms is almost pure water and from it the remaining brine can be filtered. Several successive freezing stages are employed before the brine is sufficiently concentrated to be evaporated to dryness by the application of artificial heat (ARMSTRONG and MIALL, 1946).

The concentrated brine remaining after sodium chloride has been extracted is called bitterns. These bitterns are sometimes treated to recover the materials remaining in solution. By adding calcium chloride, calcium sulphate can be precipitated which is sold as gypsum. Further concentrating causes the separation of magnesium sulphate, potash, and other salts. Magnesium chloride is finally taken from the remaining solution. Bromine is also recovered from these bitterns.

Extraction of bromine from sea water

Bromine is almost purely a marine element as over 99% of the bromine in the earth's crust is in the ocean. As indicated in Table II, sea water is about 0.0065% bromine. Bromine was discovered, around the year 1825, by a Frenchman, A. J. Balard, who found it in the bitterns obtained after the salt had been precipitated from the water of salt marshes near Montpellier. Bromine was subsequently discovered in the Stassfurt potash deposits and in the brines of wells in Michigan, Ohio, and West Virginia, in the United States. Bromine was first extracted from sea water in 1926, by treating the bitterns from the California solar salt operations. Prior to the development of the high-compression internal combustion engine, there was no demand for large quantities of bromine and the extraction of this element from well brines and from salt deposits easily supplied the market. With the development of anti-knock gasolines, a very large demand for bromine was created. In anti-knock gasolines containing tetraethyl lead, ethylene dibromide is added to prevent lead from depositing on cylinder walls, valves, and spark plug points. As the brines obtained from wells could not be pumped economically at a rate to cover this increase in demand, an alternate source was needed. Although bromine was being recovered from sea water at the time this new demand developed, it could only be recovered economically as a by-product of salt production.

While searching for additional sources of bromine, the Ethyl Corporation developed a process of precipitating bromine directly from unconcentrated sea water. In this process bromine is precipitated as the insoluble tribromoaniline by treating sea water with aniline and chlorine. To prevent hydrolysis of the chlorine, the sea water is first acidified with sulphuric acid. This process was subsequently developed on a larger scale on board a ship which had been converted into a bromine-extraction plant. Working 25 days a month, the ship was capable of producing about 75,000 lb. of bromine per month. One month's supply of reagents included 250 tons of concentrated sulphuric acid, 25 tons of aniline, and 66 tons of chlorine, stored between decks. The plant was about 70% efficient in removing bromine from sea water which contains only about 0.1 lb. per ton, a very lean ore indeed. The development work was done aboard a ship to avoid the problem of dilution of the sea water with barren effluent from the process. Later it was found that longshore currents on many coasts would work to solve the dilution problem. From a technical standpoint, the shipboard operation was successful, however, operating on the open sea with highly corrosive reagents generated a myriad of problems that could best be solved by putting the plant on land.

For a bromine-extraction facility, the site must be carefully chosen, for not only must there be no dilution of the sea water being fed to the plant by either rainfall, runoff, or by the effluent from the plant, but the sea water should be of a high and uniform salinity, of a relatively high temperature, and there must be no contamination of the sea water with organic wastes such as in sewage which would consume chlorine needlessly. At Kure Beach, North Carolina, a site was found which satisfied several of these factors, and in 1933, the Ethyl-Dow Chemical Company put into operation a 3,000 ton per year bromine-extraction facility. By 1938, the capacity of this facility had been increased to 20,000 tons per year (SHIGLEY, 1951).

In 1940, another plant was constructed at Freeport, Texas, where environmental conditions are more satisfactory for a sea-water-processing facility than they are at Kure Beach. This plant was designed for a capacity of 15,000 tons of bromine per year. In 1943, another plant of the same size was constructed at Freeport. At the end of the Second World War, the plant at Kure Beach was shut down; the facilities at Freeport now provide the United States with about 80% of her annual consumption of bromine. Fig. 8 illustrates the flow diagram for the Ethyl-Dow bromine-extraction process.

The process used at the Kure Beach plant involved mixing acid and chlorine with sea water, then feeding the mixture to the top of a brick tower packed with wood grids. The chlorine reduced the soluble bromides of sea water to a relatively volatile, elemental bromine, while the acid prevented hydrolysis of the chlorine. As the seawater-bromine mixture trickled down through the tower, air was blown upward through the mixture. The air carried the free bromine out of the sea water and thence to a soda ash absorption tower while

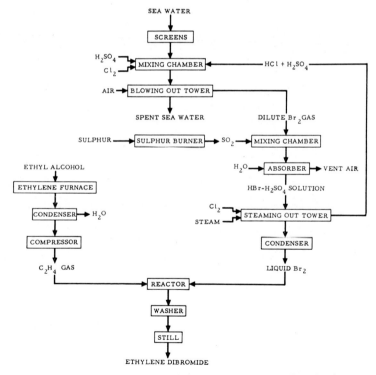

Fig. 8. Flow diagram of the Ethyl-Dow process for recovering bromine from sea water. (After SHIGLEY, 1951. Printed by permission of the *Journal of Metals*).

the bromine-free sea water was returned to the sea down stream of the plant. When the soda ash solution was saturated with bromine, the solution was treated with sulphuric acid to convert the sodium bromates and bromides to free bromine. The mixture was then pumped into a steam stripping column where the bromine was distilled off and recondensed in glass or ceramic vessels. The bromine was further refined by distillation to a product assaying 99.7% bromine.

In 1937, a modification was made to this process in which sulphur dioxide and air were used as the carrying medium in the primary bromine-stripping stage. This process delivers the bromine in the form of hydrobromic acid which allows some operating efficiencies in the subsequent refining of the bromine. Either of these processes can operate with extraction efficiencies exceeding 90%, however, the sulphur-dioxide process is now used exclusively in the United States for the direct extraction of bromine from sea water (SHIGLEY, 1951).

Magnesium from sea water

Magnesium is the lightest structural metal available. It has a specific gravity of 1.74 as opposed to aluminum with a specific gravity of 2.70 and iron with a specific gravity of 7.87. Its major application is in structural products that take advantage of its light weight such as in transportation vehicles. Magnesium is also used as an alloying agent with aluminum, in the sacrificial anodes of a cathodic protection system, for photo-flash elements, and in a variety of other ways. At present (1964) the world annually produces about 150,000 tons of this metal.

Sea water assays about 0.13% magnesium. This concentration is about one three hundredth of that of land ores of magnesium, still the major source for magnesium in the United States is sea water. Magnesium was first obtained from sea water in England (ARMSTRONG and MIALL, 1946); however, the first large scale plant for the extraction of magnesium from sea water was erected at Freeport, Texas, and put into operation early in 1941, by the Dow Chemical Company. Prior to the advent of this plant, magnesium in the United States was obtained from well brines and magnesite deposits.

There are several advantages in locating the plant at Freeport,

Texas. Cheap natural gas is available for the production of heat and electricity and the geography allows the effluent to be dumped back into the Gulf of Mexico, with little possibility of any dilution of the incoming sea water. Cheap lime is also available from calcareous shells which are mined from the floor of the Gulf a few miles from the reduction plant. The process used to recover magnesium in this plant at Freeport is illustrated in Fig. 9 while a portion of the plant itself is shown in Fig. 10.

Sea water is drawn into this plant at the rate of about a million gallons an hour through submerged gates on a canal from the Gulf to take advantage of a bottom layer of water of greater salinity than that found on the surface in the vicinity of the plant. The water is screened and continuously treated with milk of lime manufactured by calcining the oyster shells. The milk of lime reacts with the magnesium to produce an insoluble magnesium hydroxide. The resulting slurry is

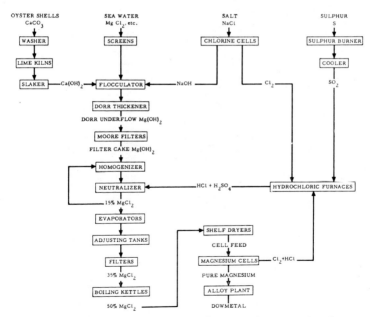

Fig. 9. Flow diagram of the Dow process for recovering magnesium from sea water. (After SHIGLEY, 1951. Printed by permission of the *Journal of Metals*).

Fig. 10. An overall view of the magnesium-processing facility at the Dow Chemical Company's Freeport, Texas, plant. In the foreground can be seen the Dorr thickeners, where the sea-water–lime mixture is pumped to allow magnesium chloride to precipitate from solution. (Photo courtesy of the Dow Chemical Company).

pumped to huge tanks where the magnesium hydroxide is allowed to settle. The slurry constitutes about 2% of the volume of the sea water originally handled, thus, in this first step of the process, a 100-fold concentration of the magnesium is effected. The effluent runs into the Brazos River and back into the Gulf at a distance point.

After filtering, the magnesium hydroxide is dissolved by hydrochloric acid yielding a magnesium-chloride solution which is concentrated by evaporation to reduce the solubility of salts carried over from the sea water. Calcium is precipitated as the insoluble calcium sulphate or gypsum by adding magnesium sulphate to the solution. After filtering to remove all the gypsum and any salt crystals that may have formed, the liquor is concentrated by evaporation. When a concentration of about 50% magnesium chloride is attained and the solution heated to a temperature of about 170°C, it is sprayed on previously dried solid $MgCl_2$. The solvent flashes into steam and the magnesium chloride is precipitated. The solid is then dried and fed into an electrolytic cell where it is decomposed to magnesium metal and chlorine gas. The chlorine is converted to hydrochloric acid and is recycled to the process. The magnesium metal is periodically drawn off the electrolytic cell and cast into ingots. The metal assays over 99.8% magnesium (SHIGLEY, 1951).

Virtually the entire United States consumption of primary magnesium metal has been obtained from sea water since the end of Second World War. During the war the U.S. Government constructed a number of magnesium-reduction plants which used magnesite, dolomite, well brines or sea water as the raw material source for the magnesium. At the end of the war, none of these plants could compete with the sea-water-extraction process even though they were granted depletion allowances and a sea water facility allowed none.

The factors involved in selecting a site for a magnesium-extraction facility are less critical than for a bromine facility unless the bromine is to be extracted in conjunction with the magnesium. Temperature of the sea water is not an important factor with a magnesium-extraction process. Only 5% of the amount of sea water is required per pound of element produced for magnesium extraction as opposed to bromine extraction. Most important is that the plant be located close by supplies of inexpensive lime, fuel, and power. Sea-water-extraction processes, in the case of magnesium, achieve recovery efficiencies of 85–90%. Higher recovery efficiencies can be attained from a technical standpoint, however, the unit capital cost per increased percent of recovery rises rapidly when efficiencies in excess of 90% are required.

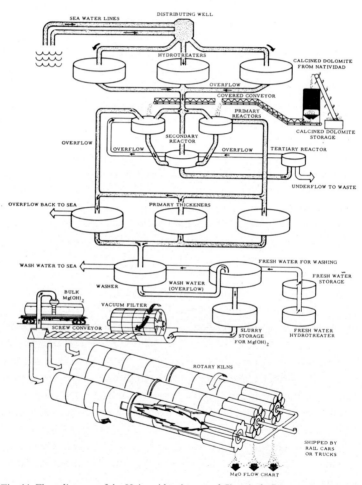

Fig. 11. Flow diagram of the Kaiser Aluminum and Chemicals Company's magnesia plant at Moss Landing, California. (Photo courtesy of the Kaiser Aluminum and Chemicals Company, Oakland, California).

The process has the inherent advantage of cheap materials handling costs as the materials in the process can be conveyed mainly by pumping. Also the process is largely continuous and is amenable to automatic process control. A distinct advantage in operating a plant like this one is the extreme uniformity of the plant feed.

Magnesium compounds

Magnesium in the form of MgO, $Mg(OH)_2$, and $MgCl_2$ finds a variety of uses as a refractory material in the lining of smelting furnaces, as pharmaceuticals, in insulations, in the manufacture of fertilizers, rayon, and paper, and in a variety of other uses. There are a number of companies around the world which produce these chemicals from sea water, notably in England and the United States. The first large-scale production of magnesium compounds from sea water was as a by-product from the bitterns remaining in the production of solar salt (SEATON, 1931; MANNING, 1936, 1938).

A process for manufacturing magnesium compounds from sea water directly is shown in Fig. 11. This process is used by the Kaiser Aluminum and Chemical Corporation at Moss Landing, California. Entering sea water is mixed with calcined dolomite to precipitate magnesium hydroxide. The hydroxide is allowed to settle in large thickeners. After settling it is taken from the thickeners, washed to remove soluble impurities and then filtered to reduce the water content to about 50%. Some of the magnesium hydroxide is sold as a homogenized filter cake for use in the manufacture of paper and magnesia insulations. The remainder of the filter cake is calcined to produce various grades of MgO which is used in various processes, producing rayon, rubber, insulations, and basic refractory brick. The Kaiser plant is shown in Fig. 12.

In the United States about 90% of the caustic-calcined magnesia and about 50% of the refractory magnesia is produced either from sea water or from well brines.

Gold from sea water

Probably more effort has been devoted to devise a method for extracting gold from sea water than any other element. Numerous

Fig. 12. Legend see p. 41.

patents have been granted involving methods and equipment for extracting gold from sea water (BARDT, 1927; BAUDIN, 1916; BAUER, 1912; CERNIK, 1926; RITTER, 1938; STOCES, 1925). In 1866, a member of the French Academy of Sciences announced that gold was found in small quantities in sea water. Around 1886, it was announced that the waters of the English Channel contained as much as 65 mg of gold per ton of water or about 0.002 ounces per ton.

At the turn of the century, Svante Arrhenius announced that the previous estimates were high by a factor of ten; however, his measurements showed at least 6 mg of gold per ton of sea water. The oceans of the world, thus, would hold some 8 trillion tons of gold. Enough gold, if it could be extracted, to make every man, woman, and child who ever lived or is living on earth a millionaire. Although several patents on processes describing the recovery of gold from sea water were issued before 1920, no gold was ever extracted.

At the end of the First World War, Dr. Fritz Haber, a brilliant German chemist and winner of a Nobel Prize, thought that the German war debt might be paid with gold extracted from the sea. Assuming that gold was in sea water in concentrations of 5–10 mg of gold per ton of sea water, Haber assembled his team, equipment, and ship to explore the oceans for the highest concentration of gold. Much to his chagrin Haber found that the concentration of gold seldom exceeded 0.001 mg of gold per ton of sea water (HABER, 1927). The highest value found in the open ocean was in the South Atlantic and it was 0.044 mg per ton. Even in San Francisco Bay which is fed by rivers flowing through gold-bearing areas, the concentration of gold does not exceed the average of the gold content of the open ocean by very much. After about 10 years of dedicated work, Haber came to the conclusion that gold could not profitably be extracted from sea water. Even today, the values obtained by Haber concerning the amount of gold in sea water are held to be somewhat inaccurate, for he apparently failed to allow for the gold content in

---

Fig. 12. An air photo of the Kaiser Aluminum and Chemical Company's magnesia-from-sea-water plant at Moss Landing, California. (Photo courtesy of Kaiser Aluminum and Chemicals Company).

the chemicals and reaction vessels he used in making his measurements.

Generally, the methods advocated for the extracting of gold from sea water involve the use of sulphide particles which have a great affinity for gold. By passing sea water over these particles, the gold is expected to adhere to the surfaces of the sulphides from which it can be later separated. Mercury was also proposed as a material for recovering gold from sea water.

Although many attempts have been made at extracting gold from sea water, there is only one instance known where any measureable amount of gold was obtained. In connection with their bromine extraction facility in North Carolina, the Dow Company became interested in investigating the possibilities of extracting other elements including gold. After processing about 15 tons of sea water, they succeeded in recovering 0.09 mg of gold, worth about U.S.$ 0.0001. To date this amount constitutes the bulk of the gold that has been extracted from sea water (TERRY, 1964).

Other materials extracted from sea water

In addition to common salt, bromine, magnesium, and magnesium compounds, several other materials are periodically extracted from sea water, generally as a by-product of salt extraction or through the agency of some plant or fish.

Iodine was first discovered in the ash of seaweed in 1811, when Bernard Courtois, a French saltpeter manufacturer, in looking for alternate sources of alkali, tried to use seaweed. In cleaning out his reaction vessels with hot concentrated sulphuric acid, he noticed a violet-colored vapor issue from the seaweed ash. The vapors condensed on the cooler parts of the vessel walls into dark metal-like crystals (ARMSTRONG and MIALL, 1946). The iodine content of some seaweeds, notably *Laminaria*, is about 0.5% on a dry-weight basis. The iodine content of sea water is about 0.05 p.p.m. or about 0.000005%. This variety of seaweed, thus, can concentrate iodine from sea water by a factor of 100,000.

Shortly after Courtois' discovery of iodine in seaweed, the medical importance of this element became known and a lively industry sprang up, mainly in northern England, extracting iodine from seaweed. By 1846, there were over twenty manufacturers of iodine from seaweed in Glasgow. Subsequent discovery of iodine in the nitrate deposits of Chile, however, put an end to the iodine-from-seaweed industry.

Potassium and sodium salts were also extracted from seaweed about this time. The processes used were not elaborate, usually simple leaching of the seaweeds with water and subsequent evaporation of the leach liquor. Another common method was to reduce the seaweed to ashes and then leach the ashes with water. The iodine generally comes out of the process as sodium or potassium iodide and is reduced to iodine by mixing with sulphuric acid and manganese dioxide.

At three different times in history seaweed served as a substantial source of minerals, once as a source of alkali, once as a source of iodine and once as a source of potash. On each occasion, however, superior methods were devised to extract these products from cheaper sources on land. Seaweed presently serves as a source of sodium alginate, an organic compound, used as a geling agent and an emulsifying agent in the preparation of numerous foodstuffs. A substantial industry, operating off the coast of Southern California, uses seaweed as a source of this chemical. Seaweed also serves as a food in many parts of the world, especially the Orient. In a number of localities near the sea it is used as a fertilizer.

Production of minerals in conjunction with the extraction of fresh water from the Ocean

Recently much interest has been devoted to the extraction of fresh water from the sea. Normally, the salts are concentrated in the effluent several times over what they are in the primary sea water. By using these brines for the extraction of minerals, several important advantages are gained; the cost of pumping is carried by the conversion plant, the brine temperature is relatively high, and the concentrations are increased by factors as high as four. If sea-water-conversion processes prove practical, the amounts of minerals available in the effluent would be, in many cases, greatly in excess of needs. For example, assuming that in a few decades we have some 100 million people living near the coasts, who consume an average of 100,000 gallons of water annually for domestic and industrial purposes. This rate of consumption would amount to about 10^{13} gallons or 10 cubic miles of water per year. If this water came from the ocean and we had a fresh water extraction efficiency of 25%, we would have the following amounts of solids passing through the conversion plants: 6.4 billion tons of sodium chloride, 240 million tons of magnesium, 160 million tons of sulphur, 800,000 tons of boron, 2,000 tons of aluminum, 400 tons of manganese, 560 tons of copper, 560 tons of uranium, 2,000 tons of molybdenum, 40 tons of silver, and about 1 ton of gold. Assuming that we could recover 10% of these minerals economically and that the people for whom the water was being produced consumed the minerals produced, the statistics of Table III indicate that only molybdenum, boron, and bromine would be produced at a rate that approximates the annual consumption of the group of people in the area concerned. Other minerals would be produced in amounts greatly exceeding the rate of consumption or in nuisance amounts.

TABLE III

TONNAGE OF MINERALS AVAILABLE FROM THE EFFLUENT BRINES OF SEVERAL SEA-WATER CONVERSION PLANTS PRODUCING 10^{13} GALLONS OF FRESH WATER PER YEAR

Material	Annual production (tons)	Per capita production for 10^8 people (tons/year)	Present U.S.A. per capita consumption (tons/year)	Ratio production/ consumption
NaCl	$64 \cdot 10^8$	64	0.145	440
Magnesium	$2.4 \cdot 10^8$	2.4	$2.5 \cdot 10^{-4}$	10,000
Sulphur	$1.6 \cdot 10^8$	1.6	0.033	50
Potassium	$68 \cdot 10^6$	0.68	0.010	68
Bromine	$1.2 \cdot 10^6$	0.012	$4.7 \cdot 10^{-4}$	25
Boron	$0.8 \cdot 10^6$	0.008	$5.5 \cdot 10^{-4}$	15
Aluminum	2,000	$2 \cdot 10^{-5}$	0.013	0.001
Manganese	400	$4 \cdot 10^{-6}$	0.0033	0.001
Copper	560	$7 \cdot 10^{-6}$	0.0067	0.001
Uranium	560	$5 \cdot 10^{-6}$	$1.4 \cdot 10^{-4}$	0.04
Molybdenum	2,000	$2 \cdot 10^{-5}$	$8.3 \cdot 10^{-5}$	24
Silver	40	$6 \cdot 10^{-7}$	$3.0 \cdot 10^{-5}$	0.02
Nickel	400	$4 \cdot 10^{-6}$	0.001	0.004
Gold	1	$2 \cdot 10^{-9}$	$5.0 \cdot 10^{-6}$	0.0004

Of course, it would not be necessary to recover all the salts, but only those that could be utilized and exhaust the remainder. In any case, because of technical difficulties, it would be unlikely that any element with a concentration less than that of boron would be recovered. An interesting note is that if all the uranium and thorium could be recovered, these elements, if used in breeder reactors, could probably provide the heat necessary to operate the conversion plant.

Large nuclear reactors have been proposed as a means of providing the heat and power for sea water conversion units (HAMMOND, 1962). The indicated production costs for the fresh water produced are about U.S.$ 0.15 per 1,000 gallons which compares favorably with the present cost of municipal water and irrigation water in certain areas. A large reactor plant could produce about 10^9 gallons per day which would be sufficient water for municipal and industrial needs for a metropolitan population of about 4 million people or would irrigate crops in an area of 500 square miles. It is unlikely, however, that such plants would be considered as a serious source of supply of water for several decades, and with the rapidity with which mineral demands and other costs change, the statistics shown in Table III are of academic value only.

ECONOMICS OF THE EXTRACTION OF MINERALS FROM SEA WATER

In studying the economics of the extraction of minerals from sea water one must consider not only the effort involved in recovering the element or material but competition from other sources of supply, distribution costs, consumption rates, and a myriad of other factors. An excellent discussion of this matter of the potential value of minerals extracted from sea water was presented by Mr. W. F. McIlhenny and Mr. D. A. Ballard of the Dow Chemical Co., in a talk before the American Chemical Society (MCILHENNY and BALLARD, 1963). Their analysis centered around the extraction unit or individual extraction facility. Because of technical difficulties in operating a larger facility, McIlhenny and Ballard chose as their extraction unit, a factory handling about 660 billion gallons of water per year (about 1.2 million gallons per min). In Table IV are presented statistics

TABLE IV

AMOUNT AND VALUE OF PRODUCTS FROM A SEA-WATER FACTORY HANDLING 660 BILLIONS GALLONS OF WATER PER YEAR

(After McIlhenny and Ballard, 1963)

Material[1] obtained	Factory production (1,000 tons/year)	Selling price (U.S.$/ton)	Product value (U.S.$ 1,000/year)	1961 U.S. consumption (1,000 tons/year)	Ratio fact. production/ consumption	Estimated land reserves at 1961 rates of consumption	
						U.S. (years)[2]	World (years)[2]
NaCl	76,300	10	763,000	26,100	2.9	+1,000	+1,000
Magnesium	45	705	31,700	45	1.0	+1,000	+1,000
Mg Compounds	5,923	53	314,000	680	5.3	+1,000	+1,000
Sulphur	2,450	24	58,800	6,000	0.4	25	NA[3]
CaSO$_4$	6,105	4	24,400	10,000	0.6	NA	NA
KCl	2,062	31	64,000	1,880	0.6	160	5,000
Br	184	430	79,000	85	2.2	+1,000	+1,000
SrSO$_4$	76	66	5,000	5	15	400	NA
Borax	113	44	5,000	100	1.1	200	NA
HF	3.8	320	1,200	330	0.01	30	50
LiOH	3.4	1,080	3,240	NA	NA	NA	NA
Iodine	0.14	2,200	306	1.3	0.11	NA	NA
MoO$_3$	0.041	3,200	132	15	$3 \cdot 10^{-3}$	150	100
Selenium	0.011	11,500	127	0.5	0.02	100	NA
U$_3$O$_8$	0.007	16,000	112	26	$3 \cdot 10^{-4}$	10	30
V$_2$O$_5$	0.014	2,760	39	6.7	$2 \cdot 10^{-3}$	NA	NA
BaSO$_4$	0.24	160	38	820	$2 \cdot 10^{-4}$	80	NA
Silver	0.0008	35,000	28	5.5	$1 \cdot 10^{-4}$	6	20
Gold	0.00002	1,000,000	20	NA	NA	NA	NA

Tin	0.008	2,240	18	49	$2 \cdot 10^{-4}$	NA	NA
Phosphates	0.22	70	15	6,100	$3 \cdot 10^{-5}$	+1,000	+1,000
Aluminum	0.03	450	13	2,300	$1 \cdot 10^{-5}$	7	120
Zinc	0.05	230	12	460	$1 \cdot 10^{-4}$	35	30
Nickel	0.006	1,580	10	180	$3 \cdot 10^{-5}$	3	40
Copper	0.008	620	5	1,230	$1 \cdot 10^{-5}$	30	35
TiO_2	0.0047	540	3	570	$1 \cdot 10^{-5}$	40	110
ThO_2	0.0002	11,000	2	0.1	$2 \cdot 10^{-3}$	+1,000	NA[3]
Cadmium	0.0004	3,600	1.5	5	$8 \cdot 10^{-5}$	6	40
As_2O_3	0.0074	100	0.7	26	$3 \cdot 10^{-4}$	100	NA
Cobalt	0.0002	3,040	0.6	5	$4 \cdot 10^{-5}$	8	100
Antimony	0.0006	650	0.4	13	$5 \cdot 10^{-5}$	3	100
MnO_2	0.005	50	0.3	1,000	$5 \cdot 10^{-6}$	NA	35

[1] Listed in form of dominant selling compound. Consumption statistics include other forms also.
[2] Assuming present level of extraction technology; +1,000 indicates material extracted from sea.
[3] NA not sufficient data available to calculate statistic.

concerning the amount of material available from such a factory, in contrast with the present rates of consumption and reserves of those materials. In this table it is indicated that only eleven minerals could be produced that would have a value of more than U.S.$ 1 million per year and only four more products would have a value greater than U.S.$ 100,000 per year.

McIlhenny and Ballard estimated that a plant to handle 1.2 million gallons of sea water per minute would cost about U.S.$ 100 million. All this plant would be required to do is take in the sea water, perform one physical or chemical manipulation per product, recover that product in a marketable form, and get rid of the effluent in a manner as not to dilute the incoming sea water. McIlhenny and Ballard estimate the operating cost of this plant by the following formula:

Operating Cost in U.S.$/year $= 1.083$ (Energy costs $+$ Raw material costs) $+ 0.1132$ (Investment costs).

The total gross value of all materials produced with an individual product value less than U.S.$ 5 million per year is about U.S.$ 7 million per year. This value is about U.S.$ 4.3 million less than the plant investment costs alone. With an extraction plant that requires the movement of the sea water through it, therefore, it would not be economic to produce any minerals or combination of minerals with a concentration in sea water less than that of boron. It might be possible to extract some of these minerals as by-products of a process recovering the other more abundant elements; however, it is very likely that the added operating costs in extracting these additional elements would exceed their value. McIlhenny and Ballard conclude, that with the existing level of technology there is little hope of commercially recovering from sea water any of the elements with a concentration less than that of boron.

NEW TECHNOLOGIES FOR MINERAL EXTRACTION FROM THE SEA

Ion exchange technology offers the possibility of using a totally different method of extracting elements from sea water, that of pulling the extraction plant, consisting of a porous container filled with ion exchange resins, through the water. Or the throughput of water could be accomplished by anchoring the resin container in some area of the ocean where ocean currents operate, such as in the Florida straits. Possibly, the resins could be painted on ships to be eluted whenever the ship visits a port with the eluting facilities, or possibly a resin container could be pulled through the ocean behind freighters. There will be major problems in the prevention of marine organisms fouling the resins and their containers, to say nothing of the problems in developing an ion exchange resin that is highly selective for some valuable element, such as uranium or silver, in the presence of a vast excess of other ions such as sodium, chloride, etc., which is the case in sea water. Great progress is being made in developing selective resins, however, and it may not be too far in the future before such resins are available.

Seaside power plants use enormous quantities of sea water for cooling purposes. In fact, a major reason for locating such plants on the sea coast is to reduce the costs involved in pumping sea water for cooling purposes. A new 300,000 kW plant to be constructed in northern California will use about 360 million gallons of water per day for cooling purposes. This water will carry with it about U.S.$ 30 million per year worth of magnesium, bromine, boron, aluminum, copper, uranium, and molybdenum, not to mention about U.S.$ 180 million worth of sodium chloride, which approximately equals the United States consumption of this material. As this plant is only one of about ten that are presently in operation along the California coast, the development of an inexpensive method of preferentially extracting one or several of the elements from the sea water in the discharge would allow recovery of commercially significant amounts of minerals. Several important advantages are available in operating with such an arrangement. The power costs in moving the sea water through the process are borne by the power plant, cheap power would be available, site selection is already made, and the intake and outlets for the power plant cooling waters are already designed to avoid mixing of the outlet waters with the inlet waters. Also, the temperature of the water is appreciably raised, which would be an advantage in some extraction processes.

Recovery of suspended matter in sea water

In addition to dissolved materials, sea water holds a great store of suspended particulate matter. Some 15% of the manganese in sea water is present in particulate form (GOLDBERG and ARRHENIUS, 1958). Most of the gold in sea water is thought to be there in the form of a colloidal suspension or as particles adhering to the surfaces of clay minerals in suspension. Much of the carbonate and silica of the sea is contained in the shells of the animals and plants which live in the water. Appreciable portions of elements such as lead and iron are also noted to be contained in sea water in particulate form (CHOW and PATTERSON, 1962; GOLDBERG, 1954).

The economics of filtering these very fine particles from sea water is anything but favorable especially if the cost of moving the water through the filtering system is carried by the recovery operation. There might be some chance of economic recovery of these materials from the outfall of a seaside power plant or other operation which uses large volumes of sea water for cooling or other purposes. Fouling of the filters by marine growths, however, would be a major stumbling block in developing such a method of recovery.

Many of these particles suspended in sea water seem to carry an electric charge. Some modification of an electrostatic process, therefore, that can operate within a conducting medium and can ignore the dissolved ions in sea water might be used to recover certain of these colloids.

The concentration of elements by marine organisms

It has been known for some time that marine organisms have a capability of effecting concentrations of certain elements in their bodies many times over the concentration of these elements in sea water. Vanadium, for example, is taken up by the mucus of certain tunicates and can be concentrated in these animals by a factor of over 280,000 times the concentration of that element is in sea water (GOLDBERG et al., 1951). Other marine organisms can effect a concentration of copper and zinc by a factor of about one million. Fish concentrate lead by a factor of 20 million in parts of their skeletons.

TABLE V

ENRICHMENT FACTORS FOR THE CONCENTRATION OF VARIOUS ELEMENTS IN MARINE ORGANISMS

Element	Concentration of element		Enrichment factor	Marine organism	Reference
	In sea water (mg/kg)	In organism[1] (mg/kg)			
Titanium	0.001	40	40,000	Algae	Walford (1958)
Vanadium	0.002	560	280,000	Tunicates	Goldberg et al. (1951)
Cobalt	0.0005	1	2,000	Algae	Walford (1958)
Nickel	0.002	5	2,500	Algae	Walford (1958)
Molybdenum	0.01	60	6,000	—	Goldberg (1957)
Iron	0.01	1,000	100,000	Algae	Walford (1958)
Lead	0.00003	700	20,000,000	Fish Bones[2]	Arrhenius et al. (1957)
Tin	0.003	1,000	330,000	Fish Bones	Arrhenius et al. (1957)
Zinc	0.01	10,000	1,000,000	Fish Bones	Arrhenius et al. (1957)
Chromium	0.00005	2	40,000	Algae	Walford (1958)
Silver	0.0003	7	21,000	—	Goldberg (1957)
Rubidium	0.12	150	1,000	Algae	Walford (1958)
Lithium	0.17	6	30	Algae	Walford (1958)
Strontium	8.0	3,000	400	Algae	Walford (1958)
Barium	0.03	100	3,300	Algae	Walford (1958)
Manganese	0.002	120	60,000	Algae	Walford (1958)
Copper	0.003	3,000	1,000,000	Fish Bones	Arrhenius et al. (1957)
Gold	0.000004	0.0014	1,400	—	Goldberg (1957)
Germanium	0.00007	0.5	7,600	—	Goldberg (1957)
Iodine	0.06	50	30,000	Algae	Walford (1958)

[1] Concentration in the dry ashes of the organisms.
[2] The element was not necessarily concentrated in the bones while the fish was living.

Table V indicates the concentration factor for various elements in marine organisms.

Although vanadium may be concentrated by a factor of 280,000 times, the resulting concentration of this element in the organism is still not sufficiently high, about 0.06% in the case of the tunicates, to regard the organisms or their remains as a possible ore of these elements. The mechanism by which these organisms can concentrate the elements, however, is important. An understanding of these processes might lead to the development of an imitative, but artificial, method of extracting and concentrating elements from very dilute solutions. Or possibly methods of breeding superconcentrators and of farming the sea with these animals and plants might be developed. Problems of farming the sea for fish or for plants, which, as sources of foodstuffs, are infinitely more valuable than they would be as sources of minerals unless some element such as gold or uranium were involved, have not been solved to any degree as yet. It is unlikely that any practical processes for winning minerals via this method could be developed in the next several decades. Marine organisms may be discovered in the future, however, which have concentrated to ore-grade in parts of their bodies certain of the more valuable elements. Silver is found to be concentrated in fish debris (ARRHENIUS et al., 1957) and seaweeds have served as a source of elements such as potassium and iodine in the past.

CHAPTER IV

THE CONTINENTAL SHELVES

The continental shelf can be defined as that area of the ocean floor between the mean low-water line and the sharp change in inclination of the ocean floor, marking the inner edge of the continental slope. This change in slope has been noted to occur at an average depth of about 430 ft. below present sea level (SHEPARD, 1963). In the past this change had been thought to occur at a depth of about 600 ft. and the 600-ft. contour line is still commonly accepted by non-oceanographers as marking the outer edge of the continental shelf. The average slope of the continental shelves of the world is less than one-eighth of one degree or about 12 ft. per mile. The average width of the continental shelves of the world is about 42 miles, but the width ranges from less than a mile to over 750 miles (SHEPARD, 1963). Fig. 13 shows the extent of the continental shelves of the world.

Continental slopes extend from the outer edge of the continental shelf to the deep ocean floor. Their average width is somewhere between 10–20 miles and their average slope is about 4°. Inclinations as steep as 45°, however, have been noted and slopes of 25° are quite common (TRUMBULL, 1958).

The continental shelves can be thought of as submerged extensions of the adjacent land areas. Their geology can generally be characterized as continuations of that on the adjacent continental areas. Although generally thought of as flat, featureless plains, the continental shelves are frequently marked with canyons, basins, and sea mounts. Glaciers, rivers, and coral growths are among the agencies that work to modify the slope and depth of the continental shelves. In some areas of the shelves, sediments are currently gathering, while in others the sediments are being eroded. In still other areas no current changes in the surficial sediment layer can be noticed.

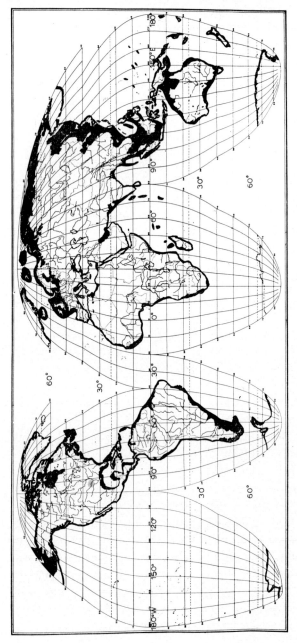

Fig. 13. Map of the world, showing the continental shelf areas in black.

Continental shelves are generally one of two types: a wide, relatively shallow and uniform plain as is generally found off stable lowland coasts, and a narrow, steep, and topographically variable area as is found along coastlines with young mountain ranges. Generally the first type of shelf is found along the east coast of the United States while the second is found along the west coast.

The distribution of sediments over the continental shelves is somewhat irregular, showing little relationship with either depth of water or distance from land. A few generalizations, however, can be made: on open shelves, sand is the most common sediment, while in protected bays and inland seas, mud is the predominant sediment. The outerpart of the open shelves generally show more coarse-grained material and bedrock than in the central or near shore areas. Offshore of large beaches, sand is normally the predominant sediment.

The rocks of the continental shelves are usually similar to those of the adjacent land; the mineral deposits of the continental shelf, therefore, can be expected to be similar also. Where explored to any great extent, as in the Gulf Coast area of southern United States or offshore California, the continental shelf areas have proved to hold mineral deposits similar to those on shore in character and approximately in quantity. These statistics are based mainly on the petroleum potentials of these two areas and not generally on the solid mineral deposits; however, in the Gulf Coast area, this observation also seems to apply for sulphur deposits in connection with salt domes.

SURFICIAL DEPOSITS OF THE CONTINENTAL SHELF

Deposits of several types of minerals on the continental shelves of the world appear to be characteristic of the environment in which they are found. Such are the phosphorite, glauconite, and calcareous shell deposits.

Calcareous shell deposits are mined in a number of areas. In the south end of San Francisco Bay, oyster shells have been mined for many years for use in the manufacture of cement and lime. From the floor of Galveston Bay in the Gulf of Mexico, the Dow Chemical Company dredges oyster shells to be used in the precipitation of magnesium from sea water. After being dredged, the oyster shells are barged to the plant at Freeport, Texas, where they are washed and stockpiled. One of the stockpiles at this plant is shown in Fig. 14. As needed, the shells are removed from the stockpiles, ground, and calcined to form CaO which is added to sea water to precipitate magnesium.

The shell deposits are found along the coasts of other Gulf Coast states and as far south along Florida as Cape Romano. Layered deposits of these shells have been found that measure over 25 ft. in thickness. Very little information exists concerning the extent of these deposits; however, they are speculated to be very large. These deposits are mined to some degree in all the Gulf Coast states. Since 1940, companies in Texas alone have mined over 45 million tons of the shells for use in various industrial processes.

Fig. 14. Oyster shells being unloaded from a barge. The shells are washed and calcined to produce quicklime which is slaked and used to precipitate magnesium from sea water. These oyster shells were dredged from Galveston and Matagorda Bays. (Photo courtesy of the Dow Chemical Company).

In many areas of the continental shelves, especially in semitropical or tropical latitudes, the offshore sediments consist mainly of pulverized coral and shell. In general, these calcareous sands could be used in the manufacture of lime or cement. Because of the low price and general availability of cement rock, however, beach or offshore sands could be expected to serve only a limited area near the deposit.

Abalone shells are gathered off California, to be used as decorative materials and in the manufacture of various souvenier items. The quantity involved, however, is small. Mother of pearl shell and other shells are taken from various tropical lagoons and bays, for use in the manufacture of jewelry and buttons. Much of this shell is hand gathered and is a major source of income for the Pacific Ocean islanders. Pearl oysters are also gathered from the continental shelves, mainly off Japan and northern Australia, or from lagoons of coral atolls.

PHOSPHORITE

A commonly occurring surficial rock of the continental shelf in certain areas of the world is phosphorite. Phosphorite, or phosphate rock, is the major industrial source of phosphorus, an element which is necessary in the life cycle of all living things. Animal life obtains this element by consuming plants, which in turn take it from the soils. If phosphorus in the form of natural or chemical fertilizers is not continually added to intensively farmed soil, that soil will soon cease to produce crops economically.

The use of phosphatic materials for fertilizers is recorded as early as 200 B.C. From that time until the middle of the 19th century, animal bones, fish and guano were the main sources phosphate fertilizers. In 1857, it was discovered that the phosphate of natural rock materials could be made soluble with acid and used as a fertilizer. Five years later, in 1862, the annual production of phosphate fertilizers, produced mainly by acidulating phosphate rock, had reached 200,000 tons in Great Britain. The production of phosphate rock has steadily increased since that time and in 1963, over 40 million tons of this material were mined for world consumption.

In the United States, about 60% of the phosphate rock mined is used in the manufacture of agricultural fertilizers. Approximately 25% is used in the manufacture of elemental phosphorus from which organic and inorganic chemicals are made. The remaining 15% of the U.S. production of phosphorite is exported to be used largely in the manufacture of fertilizers within the importing country.

The principal producers of phosphate rock are the United States, Morocco, the U.S.S.R., Tunisia, and four islands of the Pacific and Indian Oceans. World reserves of phosphate rock are about 50 billion tons. While there is no world wide shortage of phosphorite, only eight nations hold over 98% of the world's reserves. Major consumers of this vital material such as Japan, Great Britain, Germany, and Australia must import large quantities of it. All nations of the world using modern agricultural methods must manufacture and use phosphate fertilizers. As mined, phosphate rock generally contains between 10–35% of phosphorus pentoxide, P_2O_5. Most of the rock now mined is beneficiated by some means ranging from simple washing to grinding and flotation. The bulk of the marketed rock generally grades between 31–36% P_2O_5.

Sea-floor phosphorite generally contains between 20–30% P_2O_5 and has been found to be upgradable to about 32% P_2O_5. The value of the offshore phosphorite lies in the fact that, having been found off the coasts of a number of countries of the world, it can markedly reduce the delivered cost of phosphate rock to those countries. Phosphate rock is generally an inexpensive commodity; the price at the mine site being about U.S.$ 6 per short ton. Transportation costs, however, approximately double the price of delivered phosphate rock to many areas of the world.

Thus far, phosphorite has been found off Japan, off South Africa, off Argentina, along the east coast of the United States, and along the west coasts of North and South America. Very little dredging has been done off the coasts of most countries; consequently, phosphorite can be expected to be found in many other areas of the world. If found and mined, it could free many countries from the necessity of importing large quantities of phosphate rock or phosphate fertilizers.

Submarine phosphorite

Phosphatic concretions were found at several locations on the ocean floor by scientists of the "Challenger" Expedition in the 1870's. The following description of phosphorite nodules is taken from the *"Challenger" Reports* and describes nodules from the Agulhas Bank south of the Cape of Good Hope, Africa (MURRAY and RENARD, 1891). It equally well describes phosphorite found off southern California:

"They are surmounted by protuberances, penetrated by more or less profound perforations, and have, on the whole, a capricious form, being sometimes mammillated, with rounded contours, and at other times angular. Their surface has generally a glazed appearance, and is usually covered by a thin dirty-brown coating, a discoluration due to the oxides of iron and manganese. This coating, which covers all parts of the concretions, usually veils the mineralogical nature and aggregate structure... These concretions are hard and tenacious, the fundamental mass, in spite of its earthy aspect, being compact and having a hardness that does not exceed 5."

Origin of the phosphorite nodules

The *"Challenger" Reports* indicate, that the phosphorite nodules are apparently more abundant in deposits along coasts where there are great and rapid changes of temperature resulting from the meeting of cold and warm currents. In these areas, large numbers of pelagic or deep-water organisms are frequently killed by the rapid temperature changes and thus may form considerable layers of decomposing phosphatic matter on the bottom of the ocean. Large numbers of marine organisms are also killed in regions where there is a mixing of waters of different salinities, for instance, where polar and equatorial currents mingle, or where large quantities of fresh water are suddenly dumped into the ocean from floods in great rivers (MURRAY and RENARD, 1891).

In areas, where there are large amounts of decaying phosphatic matter, an atmosphere is created that allows the phosphate to dissolve in the sea water. As the dissolved phosphate migrates from this area and into the prevailing oxidizing atmosphere of the ocean, it precipitates in colloidal form. If other conditions are proper, these phosphatic colloids can agglomerate and form nodules on the sea floor. Generally colloidal precipitates of this nature possess an electric charge which causes the particles to be preferentially drawn to hard, reactive surfaces. Nodules already forming are probably an ideal surface for the further accretion of colloidal precipitates. Because there is an electrical attraction involved, the particles can adhere to the developing nodules while electrically neutral, clastic sediments are swept away.

Several other explanations of why the phosphorite nodules form in the ocean have been reviewed by DIETZ and his coworkers (1942). Generally, these explanations are extensions of theories developed to explain the formation of the Phosphoria sediments on land and none of them are without objection when applied to the offshore environment. There is probably no one simple explanation for the formation of the phosphorite nodules; the process may involve a number of factors operating in conjunction at certain periods of geologic time. For example, the phosphate could always be available in the sea water, at equilibrium with solid phases of calcium phosphates either suspended in the water or on the sea floor, until a sudden influx of gases from a volcanic eruption on the sea floor provided the fluorine to fix the phosphate in an insoluble form (MANSFIELD, 1940). In support of this hypothesis, DIETZ and coworkers (1942) have shown that sea water is generally saturated with tricalcium phosphate. It has also been noted by these authors that the Miocene, when most of the nodules seem to have been formed, was also a time of intense volcanic activity in the California area. More recent data indicate that calcite replacement by low concentrations of phosphate in sea water may be a principal mechanism of formation of these phosphorite deposits (AMES, 1959).

Recent work by Russian investigators indicates that the solubility of phosphate in sea water shows a marked dependence on the carbon-dioxide content of the sea water. Thus, when the bottom layers of water are forced to the surface in various upwellings, the resultant lowering of pressure and concomitant loss of CO_2 will result in the precipitation of phosphates (SMIRNOV, 1957).

In contrast to most of the ideas concerning the formation of sea-floor phosphorite, KRUMBEIN and GARRELS (1952) think that it probably forms in restricted anerobic basins in which the pH is relatively low. Tending to confirm this idea, is the evidence that the uranium found in sea-floor phosphorite off California exists mainly in the tetravalent state while in sea water it exists in the hexavalent state (ALTSCHULER et al., 1958). GARRELS (1960) points out that tetravalent uranium will not be stable in a marine environment in which the Eh is positive. GOLDBERG (1963a) supports these ideas using as an example a log dredged from the Gulf of Tehuantepec which was mineralized by phosphorite intrusions (GOLDBERG and PARKER, 1960). The sea water in contact with the log was depleted in oxygen and contained high concentrations of phosphate. The deficiency in oxygen results from the oxidation of the large amounts of organic matter in the sediments of this area. This idea also coincides with the observation that the phosphorite is found in areas of upwelling for these are areas of high organic productivity and consequently organic-rich sediments. It also agrees with Murray's ideas concerning the formation of phosphorite requiring an abundance of decaying phosphate-rich organic matter, presumably derived from the killing of large masses of fish or other organisms by sudden changes in temperature or salinity of the sea water (MURRAY and RENARD, 1891). In contrast to these ideas, the phosphorite off southern California is now found in some of the most highly oxidizing atmospheres of the ocean floor.

Phosphorite deposits off the coast of California

The abundant phosphorite dredged from the sea floor off the coast of California is described in some detail in several publications (CHESTERMAN, 1952; DIETZ et al., 1942; EMERY, 1960; EMERY and DIETZ, 1950; EMERY and SHEPARD, 1945; HANNA, 1952). The California phosphorite occurs as nodules which range in shape from flat slabs to irregular masses. Fig. 15 shows nodules from the southern California offshore area.

The nodules from this area are firm, dense (specific gravity of 2.62), and of medium hardness (about 5 on the Mohs scale). The surfaces of these phosphorite nodules are smooth, glazed, unweathered (indicating in situ formation), and commonly coated by a thin film of black manganese oxide. They vary in color from light brown to black. Nodules from a particular area have a group resemblance, especially in color, and range in size from oolites to chunks over a yard in diameter. The

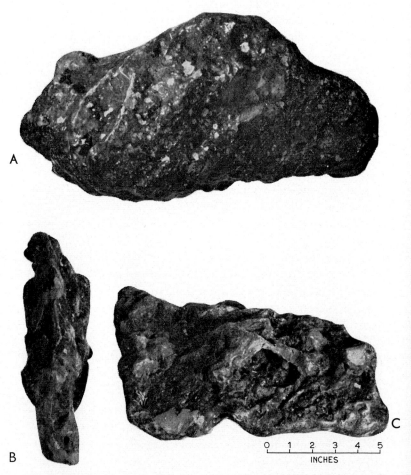

Fig. 15. Phosphorite nodules from the Forty Mile Bank deposit which is located about 40 miles due west of San Diego, California. The nodule in Fig. 15 A weighed 128 kg and was relatively free of detrital, non-phosphatic inclusions. The end and side views of the lower nodule illustrate the irregular shape many of these nodules assume. Large perforations sometimes completely penetrate the nodules.

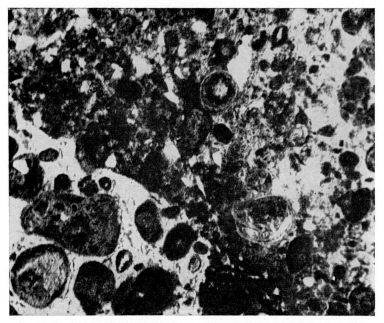

Fig. 16. A photomicrograph of a thin section of a phosphorite nodule from Forty Mile Bank, showing the oolitic structure of this material; × 50.

largest phosphorite nodule dredged off the coast of California measured 24 × 20 × 8 inches; however, larger nodules appear in sea-bottom photographs. The average diameter of all the nodules so far dredged from many different locations off California is about 2 inches. The weights of the four largest nodules thus far recovered range from 77–160 lb. (DIETZ et al., 1942).

The structure of the nodules varies considerably in different samples even within a single dredge haul. At a few stations pieces of homogeneous phosphorite were dredged, that were entirely massive collophane and francolite, having no enclosed material. The large majority of the nodules, however, are layered and many are conglomeratic. The layers are irregular and nonconcentric and vary in thickness from a few millimetres to a few centimetres. In addition to being layered, nearly all of the phosphorite is somewhat oolitic. Oolites, however, are much less abundant in these nodules than they are in the Permian

Phosphoria formation in which oolites form the mass of the rock. Fig. 16 shows the oolitic structure of a phosphorite nodule from the Forty Mile Bank deposit off San Diego.

Chemical and mineralogical composition

Petrographic examination of thin sections, supplemented by chemical, X-ray, and differential thermal analysis, shows that collophane, composed of isotropic carbonate fluorapatite, is the principal mineral in which the phosphate is present in the phosphorite nodules. Associated with the collophane is an anisotropic mineral, francolite, which is also a carbonate fluorapatite. The francolite seems to be present as a replacement of organic remains and other limy debris which were originally entrapped in the forming nodule (DIETZ et al., 1942).

Table VI lists the chemical analysis of six samples of phosphorite nodules, dredged off the California coast. Except for one sample, the nodules are surprisingly uniform in composition. The high insoluble assay of the sample from Thirty Mile Bank is due to a large physical inclusion of siliceous rock. Such inclusions are common in the phosphorite dredged from that area. The difference between 100% and the totals shown in Table VI probably represents the amount of water, MgO, and soluble SiO_2 in the phosphorite. The composition of the phosphorite nodules compares favorably with that of commercially mined phosphate rock as shown in Table VII.

The insoluble residues obtained by digesting relatively homogeneous nodules in hydrochloric acid range from 6–30% and average about 17% in the many samples tested. Clastic mineral grains and rock fragments are abundant in some residues but are not present in others. The mineral grains consist mainly of feldspars, quartz, and ferromagnesian minerals. Glauconite is relatively abundant in the nodules and occurs chiefly as rounded grains and fillings of foraminiferal tests. The insoluble residues also contain amorphous silica, carbonaceous material, and siliceous organic remains, including diatoms, radiolarians, and sponge spicules. The composition of the insoluble residues is generally similar to that of the insoluble residues of the loose sand associated with the nodules (EMERY and DIETZ, 1950).

TABLE VI

CHEMICAL COMPOSITION OF PHOSPHORITE NODULES FROM THE CALIFORNIA BORDERLAND AREA[1]

(After Dietz et al., 1942)

Constituent	Geographic location					
	Forty Mile Bank	Santa Monica Canyon	Redondo Canyon	Outer Banks	Thirty Mile Bank	Paton Escarpment
CaO	47.35	45.43	45.52	46.58	37.19	47.41
R_2O_3 [2]	0.43	0.30	2.03	0.70	3.93	1.40
P_2O_5	29.56	29.19	28.96	29.09	22.43	29.66
CO_2	3.91	4.01	4.30	4.54	4.63	4.87
F	3.31	3.12	3.07	3.15	2.47	3.36
Organic	0.10	1.90	2.25	0.44	0.35	1.50
Insoluble in HCl	2.59	3.57	4.45	3.57	20.99	2.12
Totals	87.25	87.52	90.58	88.07	91.99	90.32

[1] Compositions in weight percentages. Remaining portions largely $MgCO_3$, H_2O, and soluble SiO_2.
[2] R denotes metals.

TABLE VII

CHEMICAL COMPOSITION OF PHOSPHORITE FROM VARIOUS LOCATIONS OF THE WORLD[1]

Constituent	Sea floor		Land					
	Forty Mile Bank off California	Agulhas Bank off South Africa	Idaho	Florida	Russia	Curacao Islands	Tunisia	Morocco
CaO	47.4	37.3	48.0	36.4	27.9	50.0	44.3	51.6
R_2O_3	0.43	9.4	1.2	12.7	3.5	—	—	—
P_2O_5	29.6	22.7	32.3	31.2	17.9	37.9	29.9	32.1
CO_2	3.9	7.1	3.1	2.2	3.7	3.9	5.8	5.5
F	3.3	—	0.5	2.0	2.0	0.7	3.6	4.2
Organic	0.1	—	—	6.2	3.2	—	—	—
Totals	84.7	76.5	85.1	90.8	58.2	92.5	83.6	93.4

[1] Compositions in weight percentages. Forty Mile Bank, Idaho, Florida, and Russia assays from EMERY (1960); Agulhas Bank assay from MURRAY and RENARD (1891); Curacao Islands, Tunisia, and Morocco assays from CARO and HILL (1958).

The Foraminifera included in the structure of the nodules are predominantly Miocene in age; however, nodules with Quaternary Foraminifera have been found. EMERY and DIETZ (1950) indicate that the nodules may be forming at the present time.

Distribution

Phosphorite is found in a variety of topographic environments off the California coast. It can be obtained on the top and sides of banks, on steep escarpments which appear to be fault scarps, on the walls of

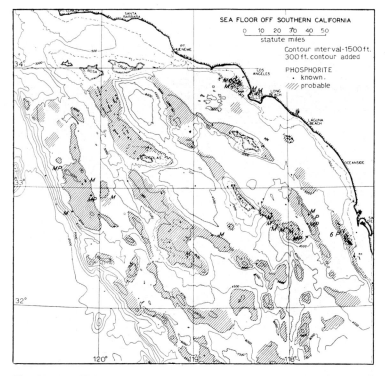

Fig. 17. Map of the sea floor off southern California, showing locations where phosphorite nodules have been recovered (dots). Areas which probably contain deposits of the phosphorite are cross-hatched. The letter M denotes samples containing Miocene Foraminifera and the letter P, Pliocene–Recent Foraminifera. (After EMERY, 1960. Printed by permission of John Wiley and Sons, Inc., New York).

submarine canyons, and on the break of the continental shelf. All of the localities where phosphorite is found are essentially non-depositional environments (DIETZ et al., 1942). Ocean-bottom currents are concentrated in these environments so that any sediment that reaches them is soon removed allowing no permanent deposition of fine sediments to take place (SHEPARD, 1941). In such environments, which exist along the offshore banks, along steep slopes, and over certain topographic highs, the slow accretion of electrically charged colloidal precipitates of phosphate may take place unimpeded by clastic sedimentation. Fig. 17 shows the stations where phosphorite has

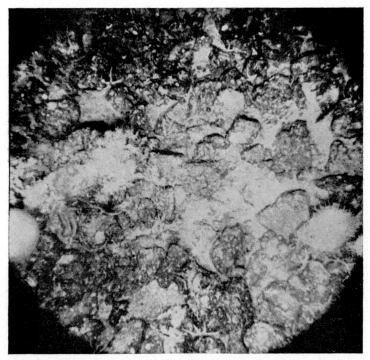

Fig. 18. A sea-floor photograph, taken at N 32°40.4′, W 118°01′, in 650 ft. of water on the northwest end of Forty Mile Bank. The area shown is 4 × 4 ft. and the estimated surface concentration is about 30 lb. of phosphorite per square foot of sea floor. (U. S. Navy Electronics Laboratory Photo).

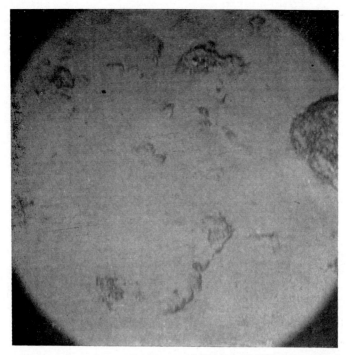

Fig. 19. A sea-floor photograph, taken at N 32°36.7′, W 117°56.5′, in 830 ft. of water on the southwest flank of Forty Mile Bank showing a total of about 80 lb. of nodules in the 16 square ft. of sea floor visible. The surficial concentration of the phosphorite nodules appears to lessen at greater depth, possibly due to the nodules being buried in clastic sediments, which can more readily accumulate in the areas lower down on the flanks of this Bank due to lessening water current velocities at greater depths. (U.S. Navy Electronics Laboratory photo).

been dredged off southern California and areas where it is suspected to be found in high concentrations on the sea floor.

Although it is conceivable that phosphorite nodules could be piled up in depth on the ocean floor in certain areas, there is, as yet, no direct evidence that the nodules are more than one layer thick. For practical economic purposes, therefore, the nodules should be assumed to exist only as a monolayer at the surface of the associated clastic sediments.

Phosphorite has been recovered from more than 125 stations off the coast of California and has been dredged in large quantities (tons) in sampling operations in at least one deposit. It undoubtedly exists at many other places where sampling operations have not yet been undertaken or where the phosphorite has been misidentified as black or brown chert or limestone. Phosphorite nodules are known at present to be found in an area extending from Point Reyes, north of San Francisco, southward to the mouth of the Gulf of California, a distance of over 1,300 miles. Within this distance, the nodules are found in deposits that lie near the inner edge of the continental slope to within a few miles of the coast.

The greatest depth from which nodules have been dredged off California is 8,400 ft., while the shallowest is about 190 ft., about 15 miles off Santa Monica. The "Challenger" Expedition recovered phosphorite nodules from a depth of 11,400 ft. at the base of the continental slope off the Cape of Good Hope (MURRAY and RENARD, 1891). The nodules from the greater depths may have been carried down the continental slopes by sediment slumps or by turbidity currents instead of forming at these depths.

Dredging and photography indicate a great abundance of phosphorite at certain locations within the southern California offshore area, such as Forty Mile Bank and Thirty Mile Bank. Fig. 18 and 19 show the phosphorite nodules on Forty Mile Bank about 40 miles due west of San Diego. Insufficient data exist to make an accurate estimate of the total tonnage of phosphorite within the California offshore area and estimates can only be considered speculative. The area in which economic recovery of phosphorite might be possible and in which phosphorite has been recovered off the California coast is roughly bounded on the north by latitude 38°, on the south by latitude 32°30′ N, on the west by the 6,000-ft. depth contour line, and on the east by the California coast line. The area within these bounds is approximately 36,000 square miles. Assuming that the phosphorite deposits cover about 10% of this area, which assumption is based on the amount of favorable topography, Table VIII indicates the quantity of phosphorite to be found at various surface concentrations within this area. MERO (1960a) made an analysis of sea-floor photographs taken at ten stations in the Forty Mile Bank deposit of phos-

TABLE VIII

ESTIMATES OF THE TONNAGE OF PHOSPHORITE NODULES AVAILABLE FOR MINING AT VARIOUS SURFACE CONCENTRATIONS WITHIN THE CALIFORNIA BORDERLAND AREA

Surface concentrations		Tonnage available for mining[1] (millions of tons)
$lb./ft.^2$	$tons/mile^2$	
1	14,000	50.4
2	28,000	101.0
5	70,000	252.0
10	140,000	504.0
20	280,000	1,010.0
30	420,000	1,512.0

[1] In 36,000 square miles of the California Borderland Area containing phosphorite nodules assuming only 10% of the area is covered with the nodules.

phorite nodules. The surface-concentration estimates on those ten photographs ranged from 5–30, and averaged 22 lb. of phosphorite nodules per square foot of sea floor. EMERY (1960) estimates that about 6,000 square miles within the southern California offshore area probably are covered by phosphorite nodules and that this area should contain about one billion tons of the nodules.

Economics of mining the California phosphorite

Although phosphorite pebbles are found in sediments in a few areas in California, notably Los Angeles and Monterey Counties, no deposit of phosphorite within the state has proved economic to mine. As a result, phosphate rock must be shipped into the state at a freight rate that doubles the cost of the mined ores at California points.

The price of phosphate rock is determined mainly by the P_2O_5 content, and in 1963, phosphate rock from Utah with a P_2O_5 content of 31% brought a price of about U.S.$ 7.30 per ton at the mine. The price of this rock at California points was about U.S.$ 15 per ton; freight charges being a little over half the delivered cost of the rock. The present rate of consumption of P_2O_5 in the state of California is about 140,000 tons annually. About 70% or some 100,000 tons is used in the manufacture of fertilizers. At an average grade of the offshore phosphorite of 29% P_2O_5, it would require about 350,000 tons annually of the nodules to supply the California market.

In a report prepared on the economics of mining phosphorite from the California offshore area, it was indicated that mining costs should range from U.S.$ 4–7 per ton of rock depending on the depth of dredging and the mining method used (MERO, 1960a). Transport costs of delivering the rock to Los Angeles or San Francisco points would be about U.S.$ 1 per ton of rock. The capital investment in an operation to mine a deposit in about 1,000 ft. of water would be about U.S.$ 3.5 million. The annual return on this investment was indicated to be about 40% after U.S. and California taxes assuming that the rock would be competitively priced on a phosphorus content basis with present sources of the material and that the annual production of rock would be about 400,000 tons.

Because of the relatively high cost of land transport in the United States, the market for the offshore phosphorite nodules would probably be limited to those places that can be reached by water transport. These markets would be the states of California, Washington, and Oregon. Mexico and Central America consume about 250,000 tons of phosphate rock annually. Japan now imports over 2 million tons of phosphate rock annually, the bulk of it coming from Florida. Certain other Asian countries, such as Formosa and the Philippines might also be substantial markets for the California phosphate rock. The consumption of phosphate rock in all these Pacific Ocean areas in 1961 was about 2.5 million tons which was valued at about U.S.$ 20 million.

Not all the phosphorite to be found in the California offshore area will be economic to mine, as some of it will be mixed with debris of a type difficult to separate from the phosphorite, some deposits will be only marginal in size, and some deposits will be located in a topographic environment from which it will be difficult to dredge with any efficiency. Probably only 10% of the deposits will be economic to mine, but there should be sufficient phosphorite in each of these deposits to make a commercial venture attractive. If 10% of the deposits will be economic to mine, there would be a reserve of about 200 years at a rate of mining of 500,000 tons per year.

World-wide sea floor phosphorite tonnages

The continental shelves of the world occupy an area of about 10 million square miles. Assuming that 10% of this shelf area contains deposits of phosphorite similar to those off Southern California, there should be in the order of $3 \cdot 10^{11}$ tons of phosphorite on the continental shelves of the world. If 10% of this amount is economic to mine, the reserves of sea-floor phosphorite would be $3 \cdot 10^{10}$ tons or about 1,000 years at the present rate of world consumption.

OTHER MINERALS ON THE CONTINENTAL SHELVES

Glauconite

An interesting authigenic mineral found in the shallower parts of the ocean, is a hydrated potassium, iron, aluminum silicate called glauconite. Widely distributed in the terrigeneous deposits of the ocean, glauconite is not commonly found in the pelagic sediments. Containing from 2–9% K_2O, glauconite could serve as a source of potash for use in agricultural fertilizers or as a source of potassium or potassium compounds.

Glauconite has been found along the coast of California in depths of water ranging from 100–300 fathoms, off the coast of South Africa in a depth of 150 fathoms, and off the east coast of Australia in about 400 fathoms. It has also been recovered off the coasts of Portugal, western Africa, eastern North America, New Zealand, the Philippines, China, Japan, Scotland, and western South America (MURRAY and RENARD, 1891).

The individual grains of glauconite found in marine muds rarely exceed 1 mm in diameter although they may occassionally be agglomerated into nodules several centimeters in diameter and cemented by a phosphatic substance (MURRAY and RENARD, 1891). The typical glauconite grains are rounded, hard, black or dark green, and often show the form and appearance of Foraminifera.

Glauconite is characteristically a constituent of green muds and sands, sometimes being the major constituent. It is also found in blue muds and in near shore *Globigerina* oozes. It is seldom found along coasts where clastic sedimentation rates are high. Conversely, it is abundant along coasts where no major rivers empty into the sea and where terrestial sedimentation rates are comparatively low. MURRAY and RENARD (1891) note that after leaving Japan, the "Challenger" Expedition did not find any glauconite until they reached the coast of South America, although they gathered hundreds of samples from the deep pelagic areas of the Pacific and from the flanks of many of the islands. Glauconite formation is apparently limited to the continental margins of the oceans. MURRAY and RENARD (1891) note that in respect to its bathymetric distribution in the ocean, glauconite appears to be most abundant about the lower limits of wave, tidal and current action, that is, in depths of about 200–300 fathoms.

Glauconite grains frequently contain traces of calcium phosphate. Also associated with glauconite are such minerals as quartz, feldspars, mica, hornblende, magnetite, and fragments of such rocks as gneiss, granite, and diabase. The constituents of glauconite apparently are derived from the weathering of these associated rocks. Glauconite is also always associated with Foraminifera or other calcareous or-

TABLE IX

CHEMICAL COMPOSITION OF GLAUCONITE FROM THE SEA FLOOR OFF THE EAST COAST OF AUSTRALIA[1]
(After MURRAY and RENARD, 1891)

Constituent	Weight percentage
K_2O	4.21
Na_2O	0.25
SiO_2	50.85
Al_2O_3	8.92
Fe_2O_3	24.40
FeO	1.66
CaO	1.26
MgO	3.13
H_2O	6.84

[1] Glauconite sample from 34°13′ S, 151°38′ E, 750 m.

ganisms, frequently filling or coating the foram tests and in some cases apparently replacing carbonate. It is very seldom found in coral muds or sands. When it is found with coral muds, the group of granitic rock minerals, mentioned previously, will also be present.

The chemical composition of a glauconitic sediment dredged from the southeast coast of Australia, is shown in Table IX. While, in relation to the present land ores of potash, glauconitic sediments could not be considered particularly rich ore, it should be noted that the cost of mining these materials should be about U.S.$ 2 per ton or less and that the glauconitic sediments can probably be concentrated by jigging, heavy media, or magnetic separation methods to a product containing 10% or so of K_2O. Because of its widespread occurrence, mineable deposits of glauconite would probably be available to most countries with coastlines where slow sedimentational rates can be found.

Little is known concerning the depth of the layer of glauconite sediments where they have been found. Off the coast of California dredge samples in such deposits have shown from 0–80% of glauconite. At depths less than 100 ft., however, glauconite rarely exceeds 1% of the total sediment. Maximum concentrations seem to occur on the outer shelf and upper continental slopes. In the southern California area, the distribution of the glauconite deposits is patchy both laterally and vertically. Although there is no direct evidence of present day deposition of glauconite in this area, PRATT (1961) feels that their almost universal association with outer shelf and bank areas which are at present receiving little or no sediment indicates that they can be forming slowly.

Assuming the layer containing the glauconite is at least several inches thick and several tens of miles in lateral extent, it should be possible to mine it at a relatively low cost. Whether the mined material can be concentrated and processed to a marketable material on an economic basis remains to be determined.

Barium sulphate concretions

Barium sulphate concretions were dredged from about 1,235 m off Colombo in the Indian Ocean in the 1880's (JONES, 1887). These concretions assayed over 75% barium sulphate. Similar barite concretions were dredged off the Kai Islands in Indonesia at a depth of 304 m and off the coast of southern California from a depth of 650 m (SVERDRUP et al., 1942). The concretions range in weight from several grams to a kilogram. They are generally irregular in shape and exhibit concentric banding (REVELLE and EMERY, 1951).

Off the south end of San Clemente Island, California, barite nodules are found to contain about 77% of barium sulphate. The cylindrical shape of many of the California concretions suggests that water containing dissolved barium moved through tubular channels in the fine-grained sea-floor sediment while the barite was precipitated when the barium came in contact with the sulphate ion of sea water. Some thin flat nodules composed of coarser grains of plagioclase, glauconite, and foraminiferal tests are indicated to have been formed by the dispersion of the barium carrying solutions through permeable layers of sea-floor sediments. Enclosed Foraminifera indicate that these concretions were formed in Pliocene sediments or possibly in Recent sediments containing reworked Pliocene forams. EMERY (1960) suggests that the baritic solutions came from magma at an unknown depth and rose along a fault plane. On contact with the sulphate-containing interstitial water of the sediments that cover the fault trace, barite was precipitated in nodular form.

Although no work has ever been done to assess the extent or concentration of these deposits, from the information available, it appears unlikely that extensive deposits of these barium-sulphate concretions will be found on the ocean floor. Such deposits that do exist, however, may be economically mineable.

Organic sediments

In various areas of the near shore regions of the oceans, basins are found which collect organic-rich sediments, that have been eroded from the adjacent continents. Because of the reducing atmosphere in these basins, resulting from a lack of adequate circulation, the organic constituents of the basin sediments are generally preserved. EMERY (1960) points out that these sediments could probably be used as fertilizers. The sediments of Santa Barbara Basin, off the coast of southern California, contain an average of about 4% of organic matter in addition to small amounts of inorganic nutrients. EMERY (1960) calculates that there should be at least $3 \cdot 10^{12}$ tons of this sediment in Santa Barbara Basin alone. A number of other basins in the southern California Borderland Area also collect organic materials. Such basins are also found in other oceans, in such areas as the fiords of Norway and in the Black Sea.

Frequently in these reducing atmosphere basins, metallic sulphides are formed. The most notable is pyrite. If springs containing copper, nickel, cobalt or other metals should empty into the floor of the basins, deposits of the sulphides of these metals would very likely form. Although no substantial deposits of metal sulphides has ever been found on the floors of these basins, very little, if any, exploration for such deposits has been done.

Sand and gravel

Probably the most important mineral commodity, from a standpoint of tonnage mined, is sand and gravel. Over half a billion tons of this material is annually produced in the United States alone. About 90% of the sand and gravel mined is used in the construction industry as aggregates in concrete or as filler material; the other 10% is marketed for glassmaking, abrasive sand, ballast, and many other uses. The bulk of the production is marketed at prices under U.S.$ 1 per ton. Because of the very low price, market areas for a specific deposit are limited, and there is practically no international trafficking whatsoever in this commodity. The consumption of sand and gravel in an area is fairly well tied to the size of the population. In many cities, the obtaining of this material can create serious problems as pits within the city limits are undesirable for a number of reasons.

As much of the population of the world is concentrated along sea coasts, the floor of the ocean is quite possibly a major future source of this material. Many beaches have been or are now being mined for sand and gravel. In a number of places, sea-floor materials are used as fill. Much of the fill for shore-side freeways in the San Francisco area has been obtained from the floor of San Francisco Bay. Long Beach harbor is being continually enlarged by impounding materials dredged from the floor of San Pedro Bay. In many cases, waves have graded deposits according to grain sizes, a relatively expensive process when it must be done on shore.

In the future, because of the inexpensive means of transport and because of the growing shortage of nearby land deposits, the sea floor may serve as a major source of sand and gravel for the construction industries in coastal areas.

Placer deposits of drowned river valleys

Many rivers which contain placer deposits empty into the sea. Certain river beds have been mined for placer minerals to the sea coast. In the recent geologic past, the level of the ocean was lowered, probably several hundred feet (KUENEN, 1950; SHEPARD, 1963). At a time of lowered sea level, rivers flowed over considerable areas of what is now the sea floor. Placer deposits can be expected to have formed along the beds of these rivers. As the ice cap melted and sea level rose flooding the land, these river valleys were submerged. Subsequent sedimentation has often completely filled the ancient river valley. Some drowned river valleys may extend as much as 300 miles to sea from the present shore line.

Some of the rivers whose channels probably extend out under the sea floor and which very likely contain placer deposits of valuable minerals are: the Salmon River of west central Alaska which contains platinum deposits and which empties into Kuskokwim Bay; Anvil Creek, near Nome, Alaska, containing gold placers; and the Orange River of South Africa which contains diamond placers. A number of rivers in the southeast Asia area contain tin placers. The valleys of many of these streams apparently extend out to sea for appreciable distances. Off Thailand and Indonesia, some of these submerged river channels are presently being mined.

In Thailand, the tin minerals are contained in alluvium submerged to a depth varying from 90–130 ft. The deposits are known to extend from the present shoreline to a point 5 miles at sea. Possibly they extend even farther offshore. Exploration of the deposits is presently accomplished by grab buckets with space control by triangulation from land bases. Mining is done with ocean-going grab dredges. A converted 5,000-ton oil tanker was outfitted with two clamshell-type buckets and has been in operation since 1957 in these deposits. Consideration is being given to a sea-going bucket ladder dredge which should considerably improve the efficiency of the operation (ROMANOWITZ, 1962). The tin minerals are concentrated on the dredging vessel by jigs and hand dressing. Waste is discharged into the sea. The concentrate is bagged and transferred by lighter from the dredge to a cargo vessel anchored nearby and then taken to a smelter in Malaya for refining.

In Indonesia, the State Mines have nine ladder-bucket dredges operating offshore in 60–100 ft. of water. Eight of these dredges are of a 14 cubic ft. capacity and one is of 9 cubic ft. capacity. The dredging is done in the open sea near the islands of Billiton, Singkep, and Bangka, off the north coast of Sumatra. The dredges are the standard bucket-ladder type. Tailings are disposed of at the rear of the dredge and the overburden and ore are worked alternately. Although there is only about 3 ft. of free board on these dredges, they ride out stormy seas anchored on the headline. Work is curtailed during the monsoon season (CRUICKSHANK, 1962).

Low frequency sonar surveys, such as those illustrated in Fig. 4, have been carried out around Billiton Island. The sea bed consists of eroded granitic rock with eminent hill and valley features. The tin minerals are concentrated in the valley sediments. By locating the valleys, therefore, most of the ground in the deposit areas can be eliminated as far as sampling operations are concerned. In the valleys the sediments may be as much as 65 ft. in thickness while along the ridges the sediments vary from 5–15 ft. in depth. The sonic surveys are periodically checked with boreholes and the correlation between the sediment boundaries as determined from the sonic data and from the actual cores has been very good. The maximum depth of reflection of the sonic waves of the method used in this area was through 150 ft. of overburden in 60 ft. of water.

Diamonds have been mined from placers in the Orange River in South Africa, at intermittent locations along its entire length. Recently, dredging was started in the ocean off the west coast of Africa near where the Orange River empties into the Atlantic. It is safe to assume that the bed of the Orange River extends out to sea for some distance and that the bed contains placer deposits of diamonds. Diamonds were apparently dispersed along the coast of Africa by surf action as they were carried into the ocean and we can expect to find diamonds dispersed along lines coinciding with the intersection of Pleistocene sea levels and coast lines.

The Salmon River in Alaska has been mined for many years for platinum. The operations are now near the coast and the placers being mined are near or below sea level (MERTIE, 1939, 1940). From all appearances, this river valley extends seaward into Kuskokwim Bay and should contain platinum deposits at least equal in grade to those being mined on shore. The shallowness of the shelf in this area indicates that the drowned river valley of the Salmon may extend several hundred miles to sea.

Methods of exploration for drowned river valleys

The first step in exploring for offshore river valley placer deposits, is a study of the onshore geology and an extrapolation of this geology to the offshore area. The location of the axis of the sea-floor valley may be immediately apparent. If so, holes can be drilled at points along the axis of the valley for sampling purposes and to determine the sediment thickness. If the location of the valley is not apparent, because of a planation of the ridges or because of total burial in sea-floor sediments of both valleys and ridges, sonic methods can generally be used to locate the valleys. Gas exploders, sparkers, or various electrical methods are used to produce a sound pulse. The sound waves penetrate the sediments as much as 2,000 ft. and a graphic recording of the reflected waves will yield a picture of the topography of the sea floor and its substructure. Fig. 20 illustrates the use of this method in locating valleys and other mineral traps buried in the sea-floor sediments. By studying the graph, distinction can be made between sediment and bedrock and sometimes even between sediment types. A systematic traversing of the offshore deposit area will yield sufficient data with which to prepare contour maps of the contact between the bedrock and the overburden. Generally, placer-type deposits are associated with magnetite and a magnetometer survey of

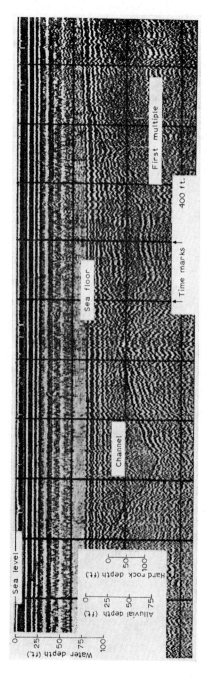

Fig. 20. Echogram of a sonic survey off the coast of Thailand, showing a buried river valley in the sea-floor sediments. Tin placers are concentrated in these valleys and the seismic profiler provides a remarkably simple and inexpensive means of locating and describing these channels. (After BECKMANN et al., 1962. Printed by permission of the *Engineering and Mining Journal*).

Fig. 21. A sea-floor-drilling vessel, the "La Cienica", operated by the Global Marine Exploration Company. Such vessels can be used to drill exploration holes in offshore beach deposits. (Photo courtesy of the Global Marine Exploration Company, Los Angeles).

the area in conjunction with the sonic survey will often help to pinpoint mineral traps within the valleys.

Sampling of the suspected deposit areas can be done by grab dredging if the sediment cover is not too thick (less than about 10 ft.) or piston coring, if the sediments are soft, or by core drilling if the sediments are compacted. In southeast Asia, the churn drill, adapted for use from a barge is commonly used. The drills are used in water depths to 50 ft. Very careful supervision of the drilling operations must be maintained to avoid contamination of the sample by side wall sluffing when the drill is ahead of the casing.

Rotary drilling has not as yet been used to any great extent in evaluating offshore mineral deposits. It is very commonly employed, of course, in the drilling of offshore exploratory and production oil wells. Methods have been developed for the drilling of such holes from floating drill rigs as shown in Fig. 21 in any depth of water and through sediment thicknesses of 20,000 ft.

CHAPTER V

STRATA UNDERLYING THE SOFT SEA-FLOOR SEDIMENTS

A cross-sectional representation of a model of the earth's crust is shown in Fig. 22. Generally, the earth's crust on the continents is composed of a surficial layer of various types of sedimentary, igneous, or metamorphic rocks. The thickness of this layer varies from 0–about 10 km. Underlying the surficial rocks is a layer of granitic material with a density of about 2.8 g per cm^3 (WORZEL and SHURBET, 1955). A basaltic layer of several kilometers thickness is thought by some authors to lie under this granitic layer (POLDERVAART, 1955). The total thickness of the crustal layers under the continents ranges from about 10 km to well over 50 km, but averages about 33 km. Underlying the crustal rocks is the mantle of the earth. The mantle is composed of rocks with a density of about 3.3 g per cm^3; these rocks

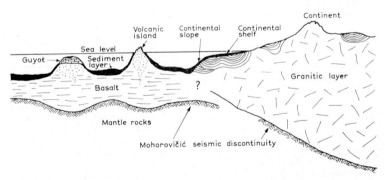

Fig. 22. A generalized cross-sectional view of the earth's crust in the area where the ocean impinges on a continent. The weight of the continental rocks depresses the mantle rocks, while the absence of such rocks in the ocean allows the Moho to rise relatively close to the earth's surface. At the Mohorovičić Seismic Discontinuity, seismic waves experience an abrupt increase in velocity indicating a marked increase in density of the mantle rocks.

constitute the bulk of the volume of the earth. The boundary between the crustal rocks and the mantle rocks is known as the Mohorovičic seismic discontinuity. It was named after a Jugoslavian seismologist, who discovered that there is a marked difference in the velocity at which seismic waves travel in mantle and crustal rocks. This boundary is commonly known as the *Moho*.

This same sequence of rock layers prevails under the continental shelves; however, sedimentary rocks generally prevail as the surficial rocks in these areas and the sequence becomes progressively thinner as the outer edge of the shelf is approached. Under the pelagic ocean, there is a marked difference in this rock sequence.

At the surface of the pelagic or far out areas of the ocean there is, of course, a layer of water which averages about 4.5 km in depth. The surficial rock layer is composed of soft, relatively unconsolidated, sediments of an average thickness of about 0.6 km (POLDERVAART, 1955). Underlying the soft sediments is a layer thought to be basaltic which averages about 7 km in thickness.

Some information concerning the chemical and mineralogical composition of these rocks can be surmised from seismic data. The angle at which sonic waves, generated usually by chemical explosions, are refracted by and the velocity at which sonic waves travel through these layers of rock can be used by seismologists to infer the density, and, thus, the composition of these layers. Seismic studies have been conducted at a number of locations on the ocean floor. In many locations, the seismic records indicate several interbedded layers of what is suspected to be consolidated pelagic sediments and basalt. It was long suspected that the second layer of the sea-floor rock column was basaltic. Without physical samples of these rocks, however, little concerning their chemical and mineralogical composition could be stated with any certitude.

Late in 1958, a project was initiated by the U.S. National Academy of Sciences, to determine the feasibility of sampling these lower layers of rock in the ocean down to and including the mantle rocks. Some preliminary drilling was done in the spring of 1961, in an area about 70 km east of Guadalupe Island, Mexico, where the water is about 3.5 km in depth. Ten holes were drilled to a maximum depth of about

180 m into the sea floor at this site. From several of these holes cores were taken of sediments which proved to be calcareous oozes (BASCOM, 1962). At the bottom of several of the holes a dark colored rock was found. Analysis of this rock showed it to be basaltic. Basalt is a fine-grained volcanic rock composed of iron, magnesium, sodium, potassium, and aluminum silicates and oxides.

Basaltic-type rocks are not notable producers of ore deposits on the continents and we have no reason to suspect that any sizeable deposits of commercially mineable minerals will be found associated with these rocks in the oceans. If deposits of valuable minerals should occur in these rocks, they would be difficult and expensive to locate and sample and, with existing technologies, next to impossible to mine profitably. Possibly, in the future, if ore deposits are located in the subsurface rocks of the deep ocean they could be exploited by some modification of the Frasch process presently used to recover sulphur; however, a chemical dissolution of the ore would have to be effected rather than a simple change of state process as in mining sulphur.

The in-situ leaching of ore deposits has, in the past, proved a difficult mining method to master. In order to have the solvent percolate through a deposit with any efficiency to contact and dissolve the ore minerals, the deposit must be permeable. Generally, if the ore deposit is permeable, the surrounding barren rock also is; consequently, there is no effective way of controlling the leaching solution and much of it is lost. With nuclear explosives, however, it may be possible to crush competent mineral-containing rock within the ocean floor (FLANGAS and SHAFFER, 1960). The surrounding unbroken rock would remain impermeable and the leaching solution could be contained. Such a method is illustrated in Fig. 23. Such methods, however, have not been developed to any great extent on the continents where a number of deposits exist on which the method could be tried. By the time all the techniques, that of drilling in deep water, setting the nuclear device properly to yield an efficient crushing of the rock in the proper zone, placing the solvent dispersing and collecting system, developing effective and inexpensive solvents, and so forth, have been mastered that will allow mining of such deposits, it is likely that synthetic mate-

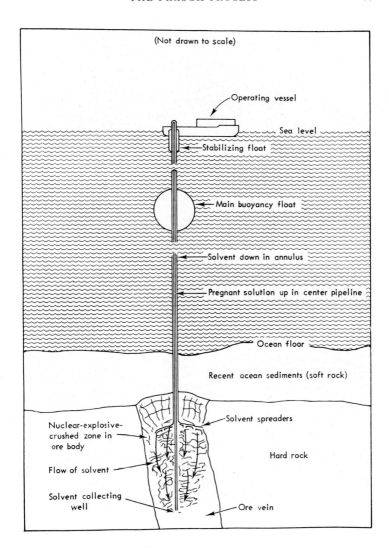

Fig. 23. Schematic diagram illustrating a proposed method of mining mineral deposits from the subsea floor. The method is similar to the Frasch process for mining sulphur; however, the ore deposit illustrated here must be broken with explosives, probably nuclear, and the minerals must be dissolved rather than merely effecting a change of state from solid to liquid as in the Frasch process.

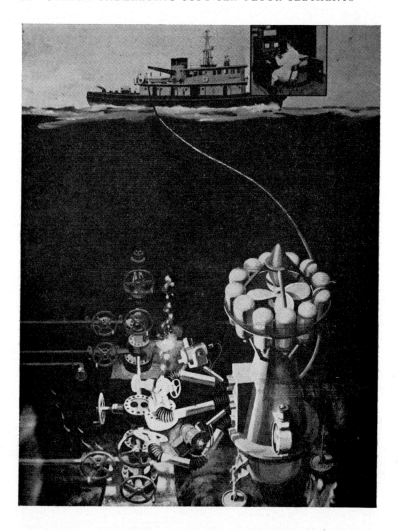

Fig. 24. Remote controlled robots have been developed for underwater use to perform operations in completing and maintaining oil well heads on the sea floor. The robots thus far developed can outperform human divers by a considerable degree. The artist's conception in this figure shows a robot tethered to a control panel on a surface vessel where the operator, via underwater television, can see all the manipulations performed by the robot. (Photo courtesy of Hughes Aircraft Company, Fullerton, California).

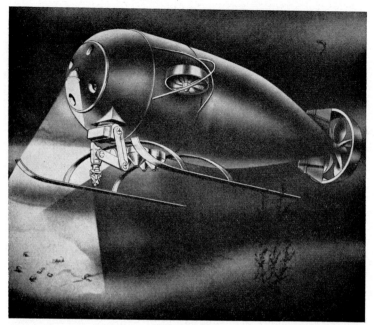

Fig. 25. Bathyscaphs are free swimming devices, which can be designed to carry men to depths in excess of 30,000 ft. below sea level. The bathyscaphe shown here, an artist's conception, consists of a two man spherical pressure chamber and an underwater manipulator, which can be used to dig in the bottom sediments and to pick objects up off the sea floor. It is powered by batteries mounted outside the pressure hull. (Photo courtesy of the General Mills Company, Minneapolis, Minnesota).

rials, made from commonly-available elements, will have reduced the need for winning minerals that may be found in any subsea-floor deposits. Possibly remote controlled robots or manned submersibles such as shown in Fig. 24 and 25 can be developed which could be used to exploit deeply submerged deposits. Shell Oil has developed one such device for use in completing sea-floor well heads and for use in operating and maintaining sea-floor oil production facilities.

DEPOSITS UNDER THE SURFICIAL SEDIMENTS OF THE CONTINENTAL SHELF

Save for its cover of water and recent sediments, the continental shelves of the world are geologically and topographically much like the continents which they border, and, in fact, can be considered an integral part of the continents. The 1953 International Committee on the Nomenclature of Ocean Bottom Features, defined the continental shelf as "the zone around the continents, extending from the low-water line to the depth at which there is a marked increase of slope to a greater depth." In general this definition is adequate; however, there are areas where two or more breaks in the slope occur before the break leading to what can be called the deep-sea floor is reached. Such breaks in slope occur off the east coast of southern United States. Thus, in addition to the break-in-slope definition of the outer edge of the continental shelf, a depth limitation is generally placed on where this break can occur. One notable attempt to define the shelf's maximum depth was at a U.N.E.S.C.O. conference at which the depth decided on was 600 m or 300 fathoms which ever was preferable by the nation concerned (SHEPARD, 1959).

The elevation of the surface of the oceans with respect to that of the continents changes continually. In the geologically recent past it has been both higher and lower than its present level. During the Ice Ages so much water was locked in the continental glaciers that sea level fell an amount generally estimated to be between 150–500 ft. (SHEPARD, 1963). At that time the continental shelves were largely uncovered and exposed to ore-deposit-forming processes which operate on the continents. Because the rocks underlying the surficial sediments of the shelves are similar to those of the adjacent continental rocks we would normally expect that the ore deposit potential of the shelf rocks would be of the same order of magnitude as the adjacent land masses.

On the continents, favorable locations for metallic mineral deposits generally occur at the contacts of igneous rocks and other dissimilar types of rock, in fissured zones, and within igneous intrusions. We can expect to find metallic mineral deposits in such areas of the continental shelf also, but the number of such deposits, relative to the total areas involved, would be expected to be much smaller in the shelves as the shelf rocks are predominantly sedimentary in character. It would also be much more difficult to locate such deposits by the geophysical, geological, or geochemical methods used to locate such deposits on land. Even the cost of drilling to confirm a deposit would be several times greater at sea as would the mining costs, if traditional metal-mining methods were to be used.

Deposits of easily soluble minerals, such as potash, that can be mined by solution mining methods on land, however, could be mined at sea at costs of about twice those on land. Greater expenses on the initial deposit can be expected until techniques are developed and refined. One such deposit in which techniques for the subsea production of minerals are being developed is the Grand Isle sulphur mine in the Gulf of Mexico. This deposit was found in the caprock of a salt dome discovered in a search for oil structures.

Salt domes are cylindrical masses of small horizontal cross section but of great vertical dimensions, which have been formed by the upward thrust of a deeply buried layer of salt. Acting under the great pressure of the overlying sediments, the salt undergoes a plastic deformation and, finding a zone of weakness, such as a fault, in the overlying sediments, is forced upward along this zone. The sediment layers through which the dome cuts are dragged upward around the

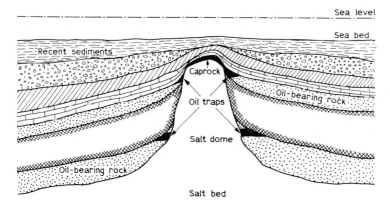

Fig. 26. A generalized, cross-sectional view of a salt dome. By pinching off the sediments through which it passes and dragging these sediment layers upward, the salt dome forms oil traps around its periphery. The doming of the overlying sediment beds also creates oil traps. The Grand Isle sulphur deposit is located in the caprock of such a salt dome.

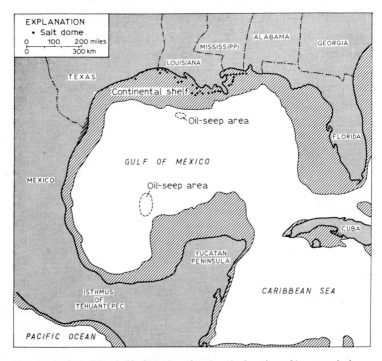

Fig. 27. A map of the Gulf of Mexico, showing the location of known salt domes and oil-seeps in the offshore area. The continental shelf is indicated by the diagonal pattern. The salt layer, providing the salt for the domes just off the Texas and Louisiana coasts, is thought to underlie a great area of the Gulf between Louisiana and the Isthmus of Tehuantepec; thus, much of the floor of the Gulf is potentially oil producing (After PEPPER, 1958).

periphery of the plug thus forming traps for accumulations of petroleum. Also, the sediment layers overlying the top of the plug are forced upward to form hemispherically-shaped domes in which petroleum can accumulate. Such a salt dome is illustrated in Fig. 26. Because the salt is of a density, that is much less than the surrounding rock, salt domes can generally be located by gravity surveys. Seismic surveys, of course, are also very useful in locating such subsurface structures. The locations of a number of such domes in the Gulf Coast area is shown in Fig. 27. One such dome was located by the Humble Oil Company under about 2,000 ft. of Gulf sediments about

DEPOSITS UNDERLYING SURFICIAL SHELF SEDIMENTS 93

7 miles seaward of Grand Isle, Louisiana. The waters of the Gulf are about 50 ft. in depth at this location. In drilling to test for oil, it was discovered that the limestone capping of the salt dome contained

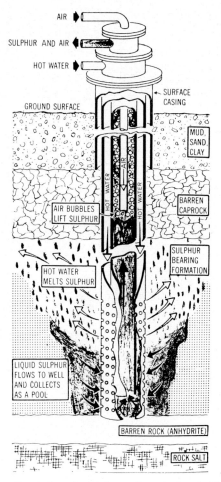

Fig. 28. A cross-sectional view of a sulphur well employing the Frasch process to remove the sulphur. Hot water is pumped down the outer pipe and out into the sulphur deposit where it melts the sulphur. Molten sulphur is forced up the middle pipe, part way to the surface. Air is pumped down the central pipe to a point below the level of the sulphur. In rising toward the surface, air bubbles carry the sulphur the rest of the way to the surface.

Fig. 29. Legend see p. 95.

considerable quantities of sulphur. The deposit covers an area of several hundred acres and varies from 220–425 ft. in thickness, making it the third largest known sulphur deposit in the United States.

The Grand Isle deposit is now being mined by the Freeport Sulphur Company, by a modification of the Frasch process in which heated sea water is used instead of fresh water. In this method, water is heated to a temperature of about 350° F and pumped under pressure into the sulphur deposit where it melts the sulphur. The molten sulphur is forced into a pipe which is placed inside the pipe which transports the water into the formation. The pressure of the water, being pumped into the deposit, forces the sulphur part way to the surface. Rather than pump the sulphur the rest of the way, air is injected into the sulphur which, in causing a decrease in the density of the sulphur column, allows it to rise the remainder of the way to the surface. The Frasch process is illustrated in Fig. 28.

At the surface, the sulphur is separated from the air and, still in a molten state, is pumped to shore in a heated submerged pipe line. In order to exploit this deposit, the largest steel structure ever to be erected in the ocean was designed and built at a cost of about U.S.$ 30,000,000. The structure, illustrated in Fig. 29, extends for a distance of 0.5 mile and rises over 60 ft. above the waters of the Gulf. It is constructed to withstand the periodic hurricanes experienced in this region. All of the necessary equipment for mining the sulphur deposit is located on this structure. Over 13,000,000 cubic ft. of gas per day are consumed in operating the water heating plant and turbogenerators. Quarters for the operating personnel are provided on the platform as well as a heliport for use in transporting men and supplies.

Fig. 29. Located about 7 miles seaward of Grand Isle, Louisiana, the Freeport Sulphur Company Grand Isle production facility is supported on the largest steel island ever built in the sea. The artificial island cost about U.S.$ 30 million and houses the heating plant, air compressors, pumps and other mining equipment as well as quarters for 120 men. At the upper ends of the Y-shaped structure are the well drilling and production platforms. Directional holes are continually drilled into new areas of the deposit as the earlier drilled areas are mined out. The platform at the left is used to drill wells to relieve the deposit of excess water. The mined sulphur is pumped to shore by a sunken, heated pipe line. (Photo courtesy of the Freeport Sulphur Company).

This deposit and various aspects of mining it are described in BAUM (1960), HANSON (1958), LEE (1960) and PALMER (1960).

Many of these salt domes have been discovered on land in the Gulf Coast region. Of the 250 domes drilled, about 10% have been found to hold sulphur deposits. About twenty of these domes have been or are being mined. We can probably expect a similar percentage of the offshore domes to contain economic sulphur deposits. It will be some time yet before it is known whether or not the Grand Isle Mine will be profitable, the destructiveness of hurricanes being a factor as yet hard to evaluate.

In addition to sulphur, common salt is also recovered from salt domes. Because of the general availability of salt from bedded sedimentary deposits and from the evaporation of sea water itself, the domes are not a particularly important source of salt. It is doubtful if the offshore domes would ever need be exploited to produce this commodity alone.

VEIN DEPOSITS

An interesting mine is in operation near Jussaro Island, about 50 miles southwest of Helsinki, Finland, where magnetite is mined from tabular veins under the Gulf of Finland. The Jussaro deposits were located by magnetic methods of geophysical prospecting. Magnetic measurements were taken at sea level from boats and on the ice during the winter. Measurements were also made with a magnetometer which could be placed on the sea floor. The magnetic anomalies were checked by sampling of the sea floor with divers and by diamond drilling from sites on rock islets in the area. A movable drilling tower supported by legs resting on the sea floor was also used in sampling of the deposits. Present detemination of reserves is made by diamond drilling from sites within the mine itself (CRUICKSHANK, 1962).

The mine was developed by driving shafts and drifts from islands in the vicinity of the veins, as indicated in Fig. 30. The veins are mined by shrink stoping, a method used in many vein-type deposits on land. Normally such a mine would be considered a land mine;

however, this mine has special problems in locating and sealing water-bearing fissures. Another mine was operated in this region through a tunnel which extended under the sea for a distance of about 2.5 km.

In Newfoundland, an iron-ore deposit, which extends out under the Atlantic Ocean, is being mined from entries on Bell Island. The reserves of ore in this deposit are measured in terms of billions of tons (PEPPER, 1958).

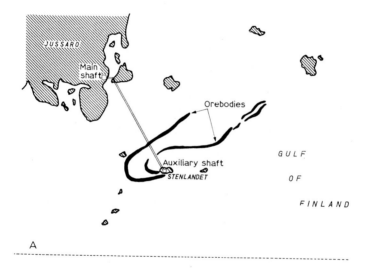

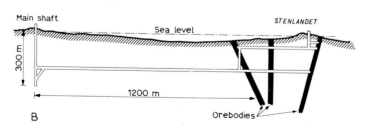

Fig. 30. The Jussaro undersea iron mine of Finland, shown in plan (A) and cross-sectional view (B).

A number of coal mines which extend out under the sea are operated in Japan, England, and Nova Scotia, through tunnels driven from shore-based shafts. As many of these deposits are exploited with normal mining methods, I would not consider them as marine resources unless they could only be discovered by exploration over or on the water and unless they must be mined with a special method because of their location under the sea.

PETROLEUM DEPOSITS

Although not generally considered a "mineral" resource, petroleum is probably the resource of greatest value and of greatest potential value found in the marine environment. Probably no industry has faced greater technical problems in dealing with the sea in the production of resources than the offshore oil industry. For these reasons, mention of this commodity is included in this volume.

Petroleum has been produced from offshore wells along the California coast since the year 1899 (THOMASSON, 1958). The offshore oil fields were discovered as extensions of onshore fields and the wells were drilled into the offshore deposits from piers or man-made islands. Some years later, a technique was perfected which permitted the drilling of directional or slanted wells into offshore fields from sites along the beach. Wells have been bottomed more than 2 miles from the coast, using this method of drilling. All these methods, drilling from islands or piers or directional drilling, of course, are severely limited to the distance from shore at which they could reach an oil deposit.

Before the Second World War, techniques had been developed to drill oil wells from floating rigs in the sheltered swamps of the Gulf Coast area of the United States. After the war, these techniques were further developed to allow the drilling of wells from such vehicles in the unsheltered offshore areas. Most of the techniques developed still depend on resting the drilling platform on the sea floor and, thus, are somewhat limited to the depth of water in which they can be used to drill a hole. Although preliminary designs have been made for bottom resting rigs to drill wells in depths of water to 600 ft., it is doubtful if such rigs will be efficient or practical to use in depths of water much more than about 300 ft. Such rigs are currently drilling in water almost 200 ft. in depth off California and one rig, designed to drill in a water depth of 250 ft., is presently under construction (ANONYMOUS, 1963a).

Fig. 31. CUSS I, the first of the floating drilling vessels, which can be used to drill exploratory or production wells in any depth of water great enough to float the vessel. This vessel was used to drill the preliminary Moholes in over 12,000 ft. of water. A few of these holes penetrated over 600 ft. into the sea-floor sediments.

During the late 1940's and early 1950's, four oil companies supported a research program designed to perfect a technique of drilling oil wells from free-floating rigs. The result of this program was the CUSS series of vessels, the first of which is shown in Fig. 31. CUSS I was capable of drilling to depths of about 15,000 ft. into the sea-floor sediments. On the preliminary phases of the Mohole Project, this vessel was used to drill about 600 ft. into the sea-floor sediments in a water depth of over 12,000 ft. A later version of this vessel, CUSS III, shown in Fig. 32 is capable of drilling in excess of 20,000 ft. into the ocean floor. While drilling wells, these vessels are anchored either by several anchors and lines splayed out from the front and aft of the vessel, or by propellers coupled with a very accurate positioning system which keeps the vessel positioned close to the center line of the

Fig. 32. The CUSS III, a later version of the CUSS I. In addition to being self-propelled, this vessel is designed to drill to depths exceeding 20,000 ft. below sea level. The heliport, on the aft end of the vessel, is used in the transportation of men and equipment to and from the vessel.

well being drilled. Such a dynamic positioning system was used in the preliminary drilling phases of the Mohole project.

In general, the mineral resources of the subsoil sections of the continental shelf are similar to those of the contiguous land, and, where oil and gas fields are found along a sea coast, extensions of those fields or additional fields can be expected to be found in similar rocks and in similar structures offshore. Off the coast of Louisiana, at least to a water depth of about 100 ft., the frequency of occurrence of major geologic structures that produce or can be expected to produce petroleum is almost exactly the same as that on the adjacent land (ATWATER, 1956).

Probably no other offshore area has been explored to the extent that the area in the Gulf Coast region of the United States has been. A common geologic structure which is found to accumulate petroleum in this area are the traps surrounding salt domes. The sediment layer providing the salt for the formation of these domes is thought to underlie a great area in the Gulf of Mexico extending from the Louisiana–Texas shore to the Isthmus of Tehuantepec, Mexico (PEPPER, 1958). Oil has also been found in southern Florida, in sediment layers at a depth of about 12,000 ft. and these sediments are suspected to extend out under the offshore area west of Florida. Oil seeps have been noted in several places (Fig. 27) outside the continental shelf limits and it is, therefore, suspected that most of the floor of the Gulf of Mexico is potentially oil producing.

The annual production of liquid hydrocarbons from the Gulf of Mexico rose from 3,500 barrels per year in 1947, to over 40,000,000 barrels per year in 1956. By 1956, the rate of well drilling had risen to about 550 wells per year of which more than 70% were oil or gas producing (THOMASSON, 1958). In mid-1963, the well-drilling rate had increased to about 750 wells per year and this rate was due to rise sharply in order to test a large number of leases that had recently been granted off Louisiana.

By 1962, the oil industry had invested over U.S.$ 2.5 billion in leases, explorations, drilling, and production facilities in the Gulf Coast offshore area. In 1963, the industry spent almost U.S.$ 500 million for leases alone off Louisiana. About 500 million barrels of petroleum have already been taken from the floor of the Gulf, the present rate of extraction exceeds 150 million barrels per year.

Through 1957, some 550 million barrels of petroleum had been produced from the area offshore California, and, in 1957, the offshore production rate was about 32 million barrels per year or about 10% of California's total petroleum production (EMERY, 1960). Although oil has been produced from the sea floor off southern California for almost 70 years, only recently has exploration been undertaken to any substantial degree along the northern California coast. Presently, exploration for oil is also underway along the Oregon and Washington coasts.

One of the great continental shelfs of the world is that off Alaska. The Alaskan shelf covers an area at least equal to the land area of Alaska including over 460,000 square miles in the Pacific Ocean. Most of the Alaskan shelf is potentially oil producing.

Outside the United States, offshore wells were producing oil at a rate of about 190 million barrels per year in 1963, and this rate is rapidly increasing. Most of this production comes from the floor of the Persian Gulf where in mid-1963, ten offshore drilling rigs were at work drilling wells in the sea floor (SWAIN, 1963).

In addition to the Persian Gulf area, offshore exploration and drilling for oil was underway along the coasts of Libya, West Africa, Nigeria, and Egypt, in Africa; Trinidad, in the West Indies; British Borneo and Sumatra, in the East Indies; and Japan. The North Sea is potentially a producer of gas and oil and extensive exploration is currently being undertaken in the area by nearly every nation that borders this body of water (SWAIN, 1963). Wells have also been drilled off the coasts of Peru, Ecuador, and Venezuela in South America.

The oil potential of the continental shelves of the world, illustrated in Fig. 11, has been estimated at 1,000 billion barrels or about equal to that of the onshore areas of the continents (PRATT, 1951).

CHAPTER VI

THE DEEP-SEA FLOOR

The dominant geologic features of the non-liquid surface of the earth are the great ocean basins. At one time thought to be generally featureless plains, these basins, with the development of the continuous sounding depth recorders, have been shown to be areas of marked relief. Well over half the deep-sea floor is covered by gently rolling hills having slopes generally not exceeding a few degrees and a relief not exceeding several hundred meters. Seamounts are commonly found on the ocean floor (any hill with a relief of more than 1 km is arbitrarily termed a seamount). MENARD (1959) estimates that there are more than 10,000 seamounts on the floor of the Pacific Ocean. Dredging has shown the preponderant proportion of the seamounts to be volcanic in origin with the origin of the remaining seamounts apparently due to fault block processes.

The ocean basins are frequently separated by great mountain ranges which periodically find surface expression as island chains. One of these ranges, the Mid-Atlantic Ridge, extends the entire length of the Atlantic Ocean. This ridge is now thought to curve around the southern tip of Africa, cross the Indian Ocean, curve around southern Australia and New Zealand and extend well up into the Pacific Ocean. Occasionally the tops of great seamounts have been planed by erosive effects. Subsequent submergence, either by a downwarping of the earth's crust under the seamount or by a relative rise in the level of the sea, results in the formation of a geomorphic feature unique to the sea floor, the guyot. First described by HESS (1946), some of the guyots contain areas approaching 5,000 km^2 on their flat upper surfaces. Bordering the island arcs and continents are the great ocean trenches, bottoming, in several cases, more than 10,500 m below the surface of the water. Fig. 33 illustrates several geomorph-

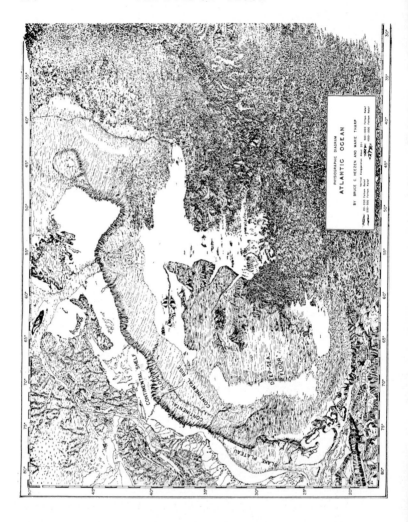

Fig. 33. A physiographic diagram of the northwestern part of the Atlantic Ocean floor, illustrating the geomorphic form of the continental shelf, slope, and rise, and the deep-sea floor. (After HEEZEN et al., 1959. Printed by permission of Dr. Bruce Heezen).

TABLE X

STATISTICS CONCERNING DEPTH ZONES IN THE OCEANS

(After SVERDRUP et al., 1942)

Areal percentages

Depth interval		Atlantic		Pacific		Indian		All oceans	
m	ft.	Individual zone	Cumulative total	Individual zone	Cumulative total	Individual zone	Cumulative total	Individual zone	Cumulative total
0–200	0–656	13.3	13.3	5.7	5.7	4.2	4.2	7.6	7.6
200–1,000	656–3,280	7.1	20.4	3.1	8.8	3.1	7.3	4.3	11.9
1,000–2,000	3,280–6,560	5.3	25.7	3.9	12.7	3.4	10.7	4.2	16.1
2,000–3,000	6,560–9,840	8.8	34.5	5.2	17.9	7.4	18.1	6.8	22.9
3,000–4,000	9,840–13,120	18.5	53.0	18.5	36.4	24.0	42.1	19.6	42.5
4,000–5,000	13,120–16,400	25.8	78.8	35.2	71.6	38.1	80.2	33.0	75.5
5,000–6,000	16,400–19,680	20.6	99.4	26.6	98.2	19.4	99.6	23.3	98.8
6,000–7,000	19,680–22,960	0.6	100.0	1.6	99.8	0.4	100.0	1.1	99.9
7,000+	22,960+	—	100.0	0.2	100.0	—	100.0	0.1	100.0

ic features of the ocean floor including the continental shelf and slope.

DEPTH OF THE OCEANS

The mean depth of the ocean is 3,800 m. Table X lists statistics concerning the percentages of the sea floor between various depths. Curiously, the deepest parts of the ocean are not in the center of the great basins, but in the trenches which border the continents and islands arcs. The greatest depth yet sounded is in the Marianas Trench where the bathyscaph, "Trieste", touched bottom in 10,850 m of water (SHEPARD, 1963) More than 98% of the ocean floor is less than 6,000 m in depth and more than 75% is less than 5,000 m in depth.

SEDIMENTS OF THE OCEAN FLOOR

Marine sediments can be subdivided into two major groups, terrigenous and pelagic. Terrigenous sediments are found near shore and generally consist of materials washed into the ocean from the continents. These sediments cover a wide range in depth, color, composition, and texture. In general, the terrigenous sediments will contain some coarse-grained material. In certain areas terrigenous sediments may be composed almost entirely of calcareous material, derived from the breakdown of the shells of various benthic animals. Terrigenous sediments may form deposits of great thickness. Seaward of the mouths of large drainage systems such as the Mississippi in the Gulf of Mexico, the sediment layers may aggregate 6,000 m in total thickness. Some of the terrigenous sediments, notably the hemipelagic ones, are named according to their color such as green mud, blue mud, yellow mud, etc. Those terrigenous sediments of the sea which contain materials of economic interest were discussed in Chapter III.

Pelagic sediments

Pelagic sediments are found in deep water, far from shore. The pelagic sediments are generally fine grained and range in color from white to a dark reddish-brown. These sediments may be either inorganic or organic in origin. Those pelagic sediments which contain less than about 30% of organic remains are called red clay. Those pelagic sediments which contain more than about 30% of organic remains are known as oozes.

The oozes are divided into two major groups, calcareous oozes, and siliceous oozes. The calcareous oozes consist predominantly of calcium carbonate in the form of the skeletal remains of various plankton animals and plants. Calcareous oozes are subdivided into three groups, *Globigerina* ooze which consists of tests of pelagic Foraminifera, pteropod ooze which consists predominantly of shells of pelagic molluscs, and coccolith ooze, which consists of the remains of a sort of oceanic Algae called Coccolithophoridae. Siliceous oozes consist largely of skeletal material produced by planktonic plants and animals. The siliceous oozes are subdivided into two groups, diatom oozes which contain large amounts of diatom frustules produced by plankton plants, and radiolarian oozes which contain large amounts of radiolarian skeletons formed by plankton animals. Fig. 34 shows the geographic distribution of the various types of sediments on the floor of the oceans and Table XI indicates the areas of the various oceans covered by these sediments. Although descriptive names are given to the various sediments of the ocean floor, it is only rarely, if ever, that a pure-type sediment is found. A few calcareous tests are always found in the purest of red clays and some clay material or other foreign matter is always found in the *Globigerina*, pteropod, diatomaceous or other oozes. Over the ocean floor, these sediment types grade into one another and it is frequently quite arbitrary whether recovered sediment is called a clay or an ooze even though standards involving the amount of organically produced material in the sediment have been set for making the classification.

Although the oozes and clays are the dominant sediments of the deep ocean floor, there are a number of other materials which, while relatively small in volume, are important because of their economic significance. These materials are manganese nodules and

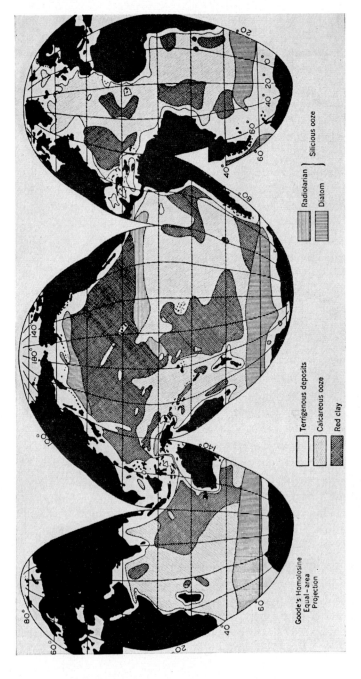

Fig. 34. A map of the world, showing the areas of the sea floor covered by the various types of sediments.

TABLE XI

DISTRIBUTION OF MARINE SEDIMENTS

(After Kuenen 1950, and Sverdrup et al., 1942)

Type of deposit	All oceans			Atlantic Ocean (millions of km^2)	Indian Ocean (millions of km^2)	Pacific Ocean (millions of km^2)
	Area covered by deposit (millions of km^2)	Sea floor areal (%)	Average depth (m)			
Terrigenous						
Shelf sediments	30	8	100			
Hemipelagic	63	18	2,300			
Pelagic	268	74	4,300			
Globigerina ooze	126	35	3,600	61.6	63.3	143.2
Pteropod ooze	2	1	2,000	40.1	34.4	51.9
Diatom ooze	31	9	3,900	2.0	—	—
Radiolarian ooze	7	2	5,300	4.1	12.6	14.4
Red clay	102	28	5,400	—	0.3	6.6
				15.9	16.0	70.3

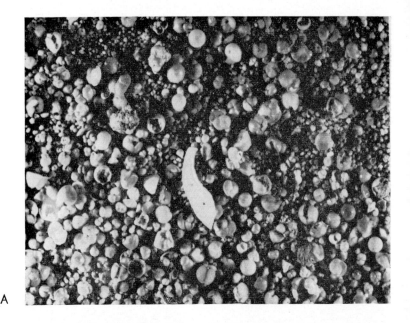

A

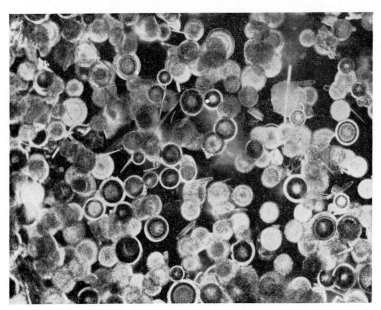

B

animal remains. All of these materials, including the clays and oozes, are of economic value and are potential sources of industrially useful minerals.

Calcareous oozes

Calcareous oozes cover some 128 million km^2 of the ocean floor, or about 36% of its total area. The average depth of the calcareous oozes is about 3,500 m, however, the oceanic depth of the surface of these deposits ranges from about 700 m to over 6,000 m. The thickness of the calcareous ooze layers in the ocean has been estimated to be about 400 m (REVELLE et al., 1955). Using this thickness and the above areal coverage it can be calculated that there are some 10^{16} tons of calcareous oozes in the oceans. These oozes are estimated to be forming at an average rate of about 1 cm per 1,000 years, thus, each year some 1.5 billions tons of calcareous ooze is added to the ocean floor. Limestone, for which these oozes could be substituted, is presently mined at an annual rate of about 0 2 billion tons per year. If only 10% of the ocean-floor deposits proved mineable, the reserves would be about 10 million years. More interesting is that the calcareous oozes, forming at an annual rate of about 1.5 billion tons, are accumulating about eight times as fast as the world is presently consuming limestone.

Of the several types of organisms which contribute to the formation of the calcareous oozes, the major contibutor, by far, is the Foraminifera, *Globigerina bulloides* (MURRAY and RENARD, 1891), and the dominant calcareous ooze on the ocean floor is termed *Globigerina* ooze. Other constituents of *Globigerina* ooze are pteropods, coccoliths, rhabdoliths, and various inorganic materials such as feldspar, magnetite, quartz, manganese-dioxide grains, and volcanic debris. These inorganic materials are distributed throughout the ocean via winds, as dust from the continental deserts, or by water currents, as pumice

---

Fig. 35. Photomicrographs, showing the shapes of the materials constituting the various sea-floor oozes. A. Calcareous ooze from N 35°06′, W 45°56′, Atlantic Ocean; × 75. B. Siliceous ooze from S 59°22′, W 142°52′, Pacific Ocean, taken with dark field illumination; × 150. (Photos courtesy of Dr. Allan Bé, Lamont Geological Observatory, Palisades, New York).

TABLE XII

CHEMICAL COMPOSITIONS OF GLOBIGERINA OOZES FROM VARIOUS OCEAN FLOOR LOCATIONS IN WEIGHT PERCENTAGES

(After Murray and Renard, 1891)

Latitude Longitude Depth (m)	N 25°52' W 19°22' 3,560	N 21°38' W 44°39' 3,480	N 21°01' W 46°29' 3,580	N 20°49' W 48°45' 4,280	S 46°46' E 45°31' 2,520	N 7°45' E 144°20' 3,400	S 38°06' W 88°02' 3,350	S 42°43' W 82°11' 2,650	S 21°15' W 14°02' 3,650
$CaCO_3$	65.2	74.5	79.2	67.6	86.4	93.1	82.6	91.3	92.5
SiO_2	18.2	11.2	9.2	17.4	6.8	1.6	4.6	2.6	1.9
Al_2O_3	5.0	5.9	3.3	6.3	2.9	1.3	5.1	1.5	1.8
Fe_2O_3	4.8					0.5	0.8	2.2	1.1
$CaSO_4$	0.7	0.5	1.2	1.9	0.8	0.3	0.6	0.7	0.2
$MgCO_3$	1.7	1.3	1.4	2.6	0.2	0.6	1.1	0.3	0.9
$Ca_3(PO_4)_2$	—	—	1.1	—	—	0.3	2.8	0.3	0.9
L.O.I.[1]	5.0	6.6	4.6	4.2	2.9	1.5	2.3	1.0	1.4

[1] L.O.I. = Loss on ignition.

TABLE XIII

STATISTICS ON DATA OF TABLE XII AND CHEMICAL COMPOSITION OF A.S.T.M. TYPES I AND II CEMENT ROCKS

Constituent	Weight percentages				
	Maximum	Minimum	Average	A.S.T.M. type I[1]	A.S.T.M. type II[1]
$CaCO_3$	92.5	65.2	81.4	74.1	73.8
SiO_2	18.2	1.6	8.2	14.1	14.6
Al_2O_3	5.1	1.3	2.9	4.2	3.2
Fe_2O_3	4.8	0.5	1.9	1.8	2.3
$CaSO_4$	1.9	0.2	0.7	—	—
$MgCO_3$	2.6	0.2	1.1	5.0	5.2

[1] Typical analyses of cement rock from TAGGART (1945).

and other debris from volcanic eruptions either under or near the oceans. The size range of *Globigerina*-ooze particles varies between 0.5–500 μ. The size distribution is relatively uniform within this size range. Fig. 35 shows a photomicrograph of various constituents of sea-floor oozes. Table XII lists the composition of samples of *Globigerina* ooze from various locations in the ocean. The calcium-carbonate content of these oozes is as high as 93%. Generally the higher calcium-carbonate contents are found in the shallower depths of the ocean which, from a mining standpoint, is a major advantage. Table XIII lists some statistics on the data of Table XII and also lists the desired analysis of A.S.T.M. Types I and II cement rock, which types constitute about 95% of the cement-rock market. As indicated by a comparison of the average analysis of the *Globigerina* ooze and the cement-rock analysis, *Globigerina* ooze would, on a compositional basis, make a good cement rock. It is high in CaO which is desirable and relatively low in magnesium, potassium, and sodium oxides, which is also desirable.

Globigerina ooze is widely distributed throughout the Atlantic Ocean, in fact, an inspection of Fig. 34 shows that the Atlantic Ocean floor consists predominantly of this type of sediment. Deposits are located within several hundred kilometers of the shores of practically all the nations bordering the Atlantic Ocean.

In the Indian Ocean about 54% of the sea floor is covered by *Globigerina* ooze. Most of the western part of the Indian Ocean is covered by calcareous oozes, while the eastern half of this ocean is covered by red clays. Calcareous ooze deposits lie within 100 km or less of the continental shores of most nations bordering this ocean.

Most of the northern half of the Pacific Ocean is covered by red clay. This type of deposit also predominates in the central part of the south Pacific, especially between the 140° and 180° lines of west longitude. *Globigerina* oozes predominate in the area bounded by Australia, New Zealand, and the 10° line of north latitude, and in the central part of the eastern south Pacific.

Globigerina ooze has a number of advantages when being considered as a substitute for limestone in the manufacture of Portland cement. In addition to its favorable composition, it is fine grained, has a large surface area, lies unconsolidated on the ocean floor, is widely available off the coasts of most maritime nations, and is available on a royalty-free basis.

The calcareous oozes could best be mined with a hydraulic dredge. With a capital investment of about U.S.$ 15 million, a dredge could be designed and built to pump about 25,000 tons of ooze per day from a depth of 4,000 m. The mining cost should be about U.S.$ 1 per ton and transport and transfer costs, from a mine site within 200 km of a port, would add about U.S.$ 1 to the cost of the ooze. As the ooze is in a finely divided state, it can be easily transferred by pumping. No crushing would be necessary in the process of converting it to Portland cement, however, a washing with fresh water would be desirable to remove salt.

In addition to the *Globigerina* oozes there are several other types of calcareous oozes. The pteropod or coccolith oozes, however, are limited in lateral extent in the ocean. In being mined for their calcium carbonate they offer no advantages over the *Globigerina* ooze other than some samples of these oozes have shown assays for $CaCO_3$ as high as 98.5% which is at least several percentage points higher than any such assays on *Globigerina* oozes. Also these other oozes generally contain less alumina and silica. Table XIV lists assays on two pteropod oozes.

TABLE XIV

CHEMICAL COMPOSITION OF PTEROPOD OOZE IN WEIGHT PERCENTAGES

(After Murray and Renard, 1891)

Latitude Longitude Depth (m)	N 18°40' W 62°56' 2,400	N 18°24' W 62°56' 820
SiO_2	4.1	2.6
Fe_2O_3	2.9	3.0
Al_2O_3	1.5	1.8
$Ca_3(PO_4)_2$	2.4	—
$CaCO_3$	80.7	84.3
$CaSO_4$	0.4	1.0
$MgCO_3$	0.7	1.8
Insoluble in HCl	3.5	2.1
H_2O	3.8	4.0

Coral sand and coral itself often show assays for $CaCO_3$ exceeding 90%. While coral deposits are used as a source of limestone on various islands, these materials possess a disadvantage in comparison with the oozes of being agglomerated into a relatively compact mass or being coarse grained and, thus, requiring a grinding step in their processing.

Siliceous oozes

Two types of siliceous oozes are found in relative abundance in the ocean floor. One type is called radiolarian ooze. It consists largely of the shells, skeletons, and spicules of Radiolaria, a plankton animal. Radiolarian oozes are found in the Indian Ocean around Cocos Island, and east of Madagascar. This type of sediment is estimated to cover some 0.3 million km² in the Indian Ocean. In the Pacific Ocean, the radiolarian ooze deposits lie along the north 10° line of latitude, extending about 5° north and south of this line and reaching from west 90° to the west 160° line of longitude. Several other patches of this ooze are found in the south central part of the Pacific. Radiolarian ooze covers an area of about 6.6 million km² in the Pacific Ocean.

Generally the radiolarian ooze lies along the borders between a calcareous ooze and red clay and is generally badly contaminated with one or the other of these bordering materials. The silica assay of the radiolarian ooze, consequently, seldom rises much above 60%. Table XV lists the assays of several of the siliceous oozes. Because of the rather high iron content of the radiolarian ooze, this material is generally some shade of red or brown. The depth of the radiolarian deposits varies between 4,300–8,200 m and averages about 5,300 m. No radiolarian oozes of note are found in the Atlantic Ocean.

Another type of siliceous sediment is diatom ooze which is composed of the frustules of planktonic plants. Deposists of this ooze are fairly well localized to the northern perimeter of the Pacific Ocean and the southern perimeters of the Pacific, Indian, and Atlantic Oceans. About 4.1 million km² of the south Atlantic Ocean floor is covered by this type of sediment, while about 12.6 km² of the south Indian Ocean, and about 14.4 km² of the Pacific Ocean is also covered with diatom ooze. The depth of these deposits ranges from 1,100 m to about 5,700 m and averages 3,900 m. In its purer forms, it is

TABLE XV

CHEMICAL COMPOSITION OF SILICEOUS OOZES IN WEIGHT PERCENTAGES

	Radiolarian ooze[1]	*Diatom ooze*[1]	*Diatom ooze*[2]
Latitude	*N 11°07'*	*S 53°55'*	*S 46°35'*
Longitude	*W 152°03'*	*E 108°35'*	*W 24°15'*
Depth (m)	5,000	3,550	4,402
SiO_2	52.9	67.9	67.4
Fe_2O_3	5.9	0.4	5.0
Al_2O_3	8.2	0.6	11.33
P_2O_5	4.0	0.0	0.10
MnO	1.7	0.0	0.19
$CaCO_3$	11.8	19.4	1.6
MgO	4.8	0.9	1.7
Na_2O	—	—	1.64
K_2O	—	—	2.15
H_2O	—	—	6.33
L.O.I.[3]	16.5	5.3	—

[1] After MURRAY and RENARD (1891).
[2] After EL WAKEEL and RILEY (1961).
[3] L.O.I. = Loss on ignition.

generally white or cream colored. At an assumed bed thickness of about 200 m, there should be about 10^{13} tons of siliceous oozes on the ocean floors. The size range of the particles of diatomaceous ooze is 1–100 μ, however, a marked maxima in the particle-size distribution histogram is found at about 10 μ (REVELLE, 1944). When it is dried diatom ooze resembles flour in color and texture.

Pure samples of diatomaceous ooze may assay in excess of 90% of SiO_2, however, there generally is a little of inorganic clay particles mixed in with these oozes. Because of the particle size differential, however, the clay particles could probably be separated from these oozes to obtain a product in excess of 99% silica on a dry-weight basis.

The uses to which this type of ooze may be put are many. It could serve in many of the ways in which diatomaceous earth is now used such as in light-weight aggregates for concrete, as a filter, in the manufacture of insulation bricks for both heat and sound, as a mineral filler, as an absorbent, and as a mild abrasive. In the United States, about half a million tons per year of diatomite are used with a mine value averaging about U.S.$ 30 per ton. While mining costs would be in the order of about U.S.$ 1 per ton for recovering this material from the sea floor, because of the long transport haul to most large consuming centers, the delivered cost would probably be in the neighborhood of about U.S.$ 10 per ton. Being fine-grained and in a nonconsolidated state, this material could be handled entirely by hydraulic pumps in transfer operations.

Animal debris

While there is little possibility of mining such objects as shark teeth, cetacean ear bones, and other animal debris found on the ocean floor some of these materials are sufficiently concentrated in certain areas of the ocean to be an important by-product in the mining of some other surficial sediment.

When a fish or other sea animal dies, the calcareous–biogenous material is generally redissolved or eaten by scavengers before it can sink to the deep-ocean floor. Shark teeth and cetacean ear bones, however, are composed of relatively large crystallites (greater than

110 Å) of tricalcium phosphate, and, consequently, are very resistant to weathering and other disintegration forces on the ocean floor. As a result, these objects can accumulate in appreciable concentrations, frequently in association with the manganese nodules for both of these materials tend to accumulate in regions where the rate of formation of the associated sediments is very low.

Shark teeth and cetacean ear bones found on the ocean floor assay as high as 34% P_2O_5 and show fluorine assays of less than 1% although, more frequently, the fluorine may assay 1–2% (Murray and Renard, 1891). Table XVI lists the composition of ear bones and shark teeth from several locations in the ocean, while Fig. 36 and 37 illustrate these objects. If the teeth or ear bones have been lying on the sea floor for any length of time, they generally have a coating of manganese dioxide and these objects show up as nuclei for manganese nodules quite often. Frequently, all or part of the interior material of the ear bones and the teeth will be replaced by manganese dioxide.

Certain sea animals have the ability to extract elements from sea

TABLE XVI

CHEMICAL COMPOSITION OF ANIMAL REMAINS ON THE SEA FLOOR IN WEIGHT PERCENTAGES

(After Murray and Renard, 1891)

	Shark teeth	Cetacean ear bone	Beak of Ziphius	Mesorostral bone of Ziphius
Latitude	S 32°36′	S 32°36′	S 33°29′	S 36°48′
Longitude	W 137°43′	W 137°43′	W 133°22′	E 19°24′
Depth (m)	4,350	4,350	4,350	3,160
SiO_2	2.5	0.6	—	—
Fe_2O_3	6.5	9.3	—	—
Al_2O_3	3.0	0.5	—	—
$Ca_3(PO_4)_2$	75.0	67.7	72.7	75.8
MnO	—	2.9	—	—
$CaCO_3$	7.5	11.0	—	—
MgO	0.7	0.3	—	3.6
F	—	—	1.7	0.03
L.O.I.[1]	4.0	4.6	3.9	—

[1] L.O.I. = Loss on ignition.

Fig. 36. Several shark teeth dredged from about 2,500 ft. of water about 100 miles off Jacksonville, Florida. The tooth in Fig. 36A is serving as the nucleus of a manganese nodule. The surfaces of these teeth were highly polished with a varnish-like coating. The interior parts of the teeth had largely been replaced by manganese and iron oxides.

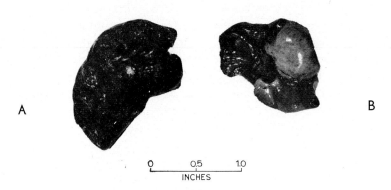

Fig. 37. Cetacean ear bones dredged from the Blake Plateau, off the east coast of the United States. Like shark teeth dredged from this area, the ear bones have a polished, varnish-like coating.

water and concentrate these elements in various parts of their bodies. Some of the tunicates can concentrate vanadium in their bodies to over 50,000 times that of the concentration of vanadium in sea water. Analyses performed by ARRHENIUS et al. (1957) have shown several percent of rare-earth elements, 0.6–1.5% of zinc, 0.1–0.5% of copper, 0.05–0.15% of tin, and 0.03–0.1% of lead in fish skeletal remains. Concentrations of nickel and silver are also found in fish debris. Very little data exist at the present time concerning the extent of fish-debris deposits on the ocean floor. It is possible, although probably very unlikely, that fish remains could collect in certain basins in the ocean floor in areas of large fish populations, more rapidly than they disintegrate. Such deposits, if they exist, however, would be expected to be quite limited in volume and probably found only in relatively shallow water, probably on the continental shelves.

Other minerals

Zeolites are frequently found on the ocean floor and various metals, such as germanium, are sometimes found concentrated in certain zeolitic minerals (WARDANI, 1959). Very little is known about the concentration of such zeolites or their distribution on the sea floor.

Phillipsite is a common mineral of some pelagic clays, constituting, in some cases, 50% of the clay (BRAMLETTE, 1961). Although crystals of phillipsite have been found to be most abundant in those areas farthest removed from the continents such as the south-central part of the Pacific Ocean or the central part of the Indian Ocean, there seems to be no correlation with sediment type and the appearance of this mineral. The association of abundant crystals of phillipsite with the red clays is probably due to the slow rate of formation of the clays. Table XVII shows the chemical composition of phillipsite crystals dredged from 4,800 m about 600 km north of Tahiti. MURRAY and RENARD (1891) note that phillipsite is always found in association with volcanic debris and they conclude that phillipsite and other zeolites are formed by the reprecipitation of chemical constituents very recently weathered from the volcanic debris. The transfer process is indicated to take place possibly within the upper layer of the sediments as well as at the surface of these sediments. The crystals of phillipsite produced are rarely of a size greater than about 0.01 mm in diameter. They frequently show crossed twins.

Palagonite, the end product in the weathering and hydration of basaltic-volcanic glasses, also is abundant in the red-clay areas of the ocean. Frequently, this material serves as a nucleus for the formation of a manganese nodule. As shown in Table XVII, both the phillipsite and palagonite sediments contain appreciable amounts of potash. While it is doubted, because of the great depth in which they are found, that these materials would be of sufficient value to mine from the ocean floor by themselves, it is very possible that they could be recovered as a by-product in the mining of manganese nodules or other materials from the deep-sea floor.

Cosmic spherules are frequently found in the pelagic sediments. Often these spherules are composed of nickel and iron. Because of the sparseness of these materials, however, it is doubtful that they could even be recovered on a commercial basis as a by-product in the production of some other material.

TABLE XVII

CHEMICAL COMPOSITION OF OCEAN FLOOR PHILLIPSITE AND PALAGONITE IN WEIGHT PERCENTAGES

(After MURRAY and RENARD, 1891)

	Phillipsite	*Palagonite*
Latitude	*S 11°20'*	*S 13°28'*
Longitude	*W 150°30'*	*W 149°30'*
Depth (m)	4,770	4,300
SiO_2	49.9	44.7
Fe_2O_3	5.5	14.6
Al_2O_3	16.5	16.3
MnO	0.4	2.9
CaO	1.4	1.9
MgO	1.2	2.2
K_2O	5.1	4.0
Na_2O	4.6	4.5
H_2O	16.7	9.6

Red clay

The hydrated-aluminum-silicate clay-mineral is generally the end product in the weathering of igneous rocks. Such weathered products are generally very finely divided and when mixed with water can remain in suspension almost indefinitely. Dumped into the ocean by rivers and streams, these clay particles can float to any part of the ocean before traversing the water column to come to rest on the ocean floor. There, mixed with iron oxides in a generally highly oxidizing atmosphere, the clay deposits take on a characteristic red color and derive their name, red clay. It is a good descriptive name and appropriately applies in the bulk of pelagic clay samples. The red coloration, however, is frequently modified by the inclusion of manganese-peroxide grains, which give the clay an overall dark brown color, or by calcareous or siliceous oozes which lighten the reddish color.

Like most clays, pelagic red clay is soft and plastic. It has the consistency of a heavy grease. When dried, it generally contracts and solidifies into a hard compact mass. In this state, the red clays possess considerable engineering strength. The red clays frequently exhibit a gritty feeling when squeezed between the fingers. This gritty feeling is due to the inclusion in the clay matrix of manganese-dioxide grains, cosmic spherules, zeolite crystals, foram tests, shards of pumice or other minerals, and rock fragments. In certain areas, such as off the west coast of North Africa or off western Australia, winds carry large quantities of dust from desert areas great distances to sea and the sediments in these areas are heavily contaminated with these wind-borne products. Debris from volcanic eruptions such as dust or pumice can be distributed throughout the ocean by wind and water currents in a relatively short time and these materials are frequently found as constituents of the pelagic sediments. Particles exceeding 0.1 mm in diameter, however, seldom constitute more than 1–2% of the weight of the red clays save for those instances where very large (several centimeters and sometimes even meters in diameter) blocks of pumice, or ice or seaweed-rafted boulders are found at the surface of the sediments. Weathering of the pumice and other volcanic debris was thought by Murray (MURRAY and RENARD, 1891) to play an important part in providing the aluminum silicates for clay formation in situ.

As part of the operations aboard the H. M. S. "Challenger" during the 1873–76 cruise, a portion of all sediment samples was screened and Table XVIII shows the average results obtained in processing about 60 samples of red-clay sediments. The "minerals" constituent consisted of such materials as magnetite, manganese grains, hornblende, palagonite, quartz, plagioclase, mica, zeolites, cosmic spherules, rock fragments, zircon, and tourmaline. Table XIX shows the chemical composition of red clays from the various oceans and from various locations in the Pacific Ocean.

While, from a mineral-resource standpoint, the composition of the red clay is not particularly exciting, this material may have some value as a raw material to be used in the manufacture of clay products or it may, in the future, serve as a source of various metals. While the average assay for Al_2O_3 is about 15%, certain of the samples assayed in excess of 25% Al_2O_3, which places the red clay at about the same compositional level as some of the continental clays that are being considered as a source of aluminum. Copper, nickel, cobalt, vanadium, lead, zirconium, and several of the rare earths show up in amounts of several hundredths or several tenths of a percent.

TABLE XVIII

CONSTITUENTS OF PELAGIC RED CLAY

(After MURRAY and RENARD, 1891)

Constituent	*Weight percentage*
Pelagic Foraminifera	4.8
Bottom-living Foraminifera	0.6
Other organisms	1.3
Siliceous organisms	2.4
Minerals	5.6
Fine washings[1]	85.4
Total	100.1

[1] Mainly hydrated aluminum-silicates but also containing particles of all other constituents.

TABLE XIX

CHEMICAL COMPOSITION OF RED CLAYS FROM VARIOUS PACIFIC LOCATIONS IN WEIGHT PERCENTAGES

(After Goldberg and Arrhenius, 1958)

Latitude / Longitude / Depth (m)	S 16°36' W 162°43' 5,125	S 12°46' W 143°33' 4,380	N 9°17' W 124°09' 4,410	N 19°01' W 177°19' 4,774	N 27°38' W 124°26' 4,400	N 35°09' W 157°17' 5,600	N 53°01' W 176°15' 3,660	Composite of 51 samples[1]	Atlantic Ocean sample[2]
SiO_2	45.8	47.0	56.0	61.3	57.5	52.8	67.0	54.5	53.3
Al_2O_3	20.5	14.7	15.9	19.5	17.8	14.8	11.4	15.9	23.7
Fe	6.2	6.2	4.6	3.5	4.6	5.7	4.0	6.7	5.1
Ti	0.78	0.36	0.38	0.43	0.49	0.44	0.33	0.6	0.6
Mg	1.8	1.9	1.7	2.4	2.4	2.2	2.3	2.0	2.1
Ca	3.3	5.9	2.0	5.8	1.2	2.1	2.9	1.4	3.6
Na	4.5	4.4	3.9	4.1	2.8	2.6	2.6	1.6	2.8
K	2.6	2.2	2.5	3.1	2.2	2.9	1.1	2.4	2.6
Sr	0.05	0.041	0.035	0.061	0.040	0.030	0.036	0.047	0.019
Ba	0.069	0.26	1.2	0.16	0.60	0.11	0.10	0.18	0.45
Mn	1.5	3.0	0.87	1.6	0.46	1.6	0.14	0.7	0.09
Ni	0.028	0.039	0.031	0.083	0.011	0.026	0.003	0.02	0.012
Cu	0.077	0.14	0.20	0.093	0.066	0.038	0.010	0.019	0.012
Co	0.031	0.024	0.011	0.031	0.009	0.009	0.001	0.01	0.008
Cr	0.006	0.014	0.004	0.005	0.010	0.008	0.006	0.008	0.037
V	0.043	0.020	0.012	0.025	0.031	0.021	0.014	0.024	0.018
Pb	0.017	0.021	0.012	0.008	—	0.010	0.012	0.007	0.008
Mo	0.0065	0.0008	0.0017	0.011	0.037	—	—	—	—
Zr	0.021	0.013	0.012	0.017	0.018	0.023	0.010	—	0.03
Y	0.016	0.039	0.013	0.0038	0.005	0.017	0.002	—	—
Sc	0.0018	0.0044	0.0037	0.0040	0.0022	0.0028	0.0020	—	—

[1] After Sverdrup et al. (1942).
[2] After Correns (1939).

Copper assays as high as 0.20% in some red clays. GOLDBERG and ARRHENIUS (1958) list analyses for some 26 elements of clays from various locations in the Pacific Ocean and, in general, the high copper, nickel, cobalt, etc., assays could be correlated with a rather high manganese content. It is safe to assume, therefore, that these minor metals are largely associated with the manganese grains which are found widespread throughout the red clays. If some means could be devised to screen these small particles of manganese dioxide which average about 0.1 mm in diameter from the red clay, the oversize would probably be an economic resource of these metals. The manganese content of the clays is about 1–2%, therefore, a concentrating factor of about 25 would be involved in such a process. From an economic standpoint, the major disadvantage is that 25 tons of material must be lifted from the sea floor for each ton of useful material obtained. Unless some means of making the separation at the sea floor could be devised, therefore, it is doubtful that the red clays would be economic to mine for the manganese grains and their associated metals. If the red clays were to be mined for some other material such as aluminum, or for some other purpose such as use in the manufacture of construction materials, it might be possible to recover the manganese grains as a by-product.

In any mining venture, of course, the highest grade of clay would be mined first, other conditions being equal. The percentage of manganese grains in some of the red clays approximates 5% and it is quite possible that this percentage may be appreciably higher in unsampled areas of the ocean floor. Because of the very low costs in moving the materials from the sea floor to the surface (probably in the neighborhood of U.S.$ 1 per ton or less for the red clays) it would become attractive to mine the red clays when the manganese-grain content approached 10%, assuming, of course, that a simple and inexpensive means of removing the grains from the rest of the clay can be devised.

About half of the Pacific Ocean is covered with red-clay sediments. They predominate in the north Pacific and in the south-central Pacific. Red clay is estimated to cover some 70 million km^2 in this ocean. About 25% of the ocean floor in both the Atlantic and Indian

TABLE XX

STATISTICS ON AMOUNT OF AND RATE OF ACCUMULATION OF VARIOUS ELEMENTS IN RED CLAY[1]

Element	Abundance in red clay[2] (weight percent)	Amount in red clay (trillions of tons)	Rate of accumulation in red clay (millions of tons/year)	World[3] rate of consumption (millions of tons/year)	Ratio $\dfrac{\text{Amount in red clay}}{\text{Annual consumption}}$ ($\times 10^6$)	Ratio $\dfrac{\text{Rate of accumulation}}{\text{Rate of consumption}}$	World[4] reserves in 1958 (millions of tons)	Ratio $\dfrac{\text{amount in red clay}}{\text{world reserves}}$ ($\times 10^3$)
Al	9.2	920.0	46.0	4.72	200.0	10	570	1,620
Mn	1.25	125.0	6.3	6.7	19.0	1	320	390
Ti	0.73	73.0	3.7	1.3	56.0	3	140	520
V	0.045	4.5	0.23	0.008	550.0	28	NA[5]	—
Fe	6.5	650.0	32.5	262.5[6]	2.5	0.1	1,350	480
Co	0.016	1.6	0.08	0.015	110.0	5	1.6	1,000
Ni	0.032	3.2	0.16	0.36	8.9	0.5	13.5	220
Cu	0.074	7.4	0.37	4.6	1.6	0.1	150	50
Zr	0.018	1.8	0.09	0.002	900.0	45.0	NA	—
Pb	0.015	1.5	0.08	2.4	0.6	0.03	43	35
Mo	0.0045	0.45	0.023	0.040	11.0	0.6	3	150

[1] Based on an oceanic tonnage of red clay of 10^{16} tons and a rate of accumulation of $5 \cdot 10^8$ tons per year. All quantities expressed in metric tons.
[2] After GOLDBERG and ARRHENIUS (1958).
[3] From *Encyclopaedia Britannica Book of the Year* (1963).
[4] After McILHENNY and BALLARD (1963).
[5] No data available with which to calculate statistic.
[6] Primary iron.

Oceans are covered with red clay. The total coverage in all the oceans is about 102.2 million km². At an average thickness of about 200 m (REVELLE et al., 1955) there should be some 10^{16} tons of red clay on the ocean floors. At an average rate of formation of 5 mm per 1,000 years, the annual rate of accumulation of the red clays is about $5 \cdot 10^8$ tons. Table XX lists some statistics concerning the amount of various elements in the red clay presently on the ocean floor and the rate at which the elements are annually accumulating in the red clay.

MANGANESE NODULES

One of the most interesting discoveries made by the "Challenger" Expedition (1873–76) was that of the abundance of black, hydrous, manganese-dioxide concretions on the floors of the three major oceans. At the turn of this century, Alexander Agassiz of Harvard University, on a number of oceanographic expeditions in the eastern Pacific Ocean, recovered nodules at almost every station in deep water at which he dredged. Manganese nodules and crusts are a form of pelagic sedimentation, but because of their relatively small volume are of minor importance when considering the totality of oceanic sediments. From an economic standpoint, they are the most important sediment of the deep-ocean floor.

Physical forms of ocean-floor manganese–iron oxides

Manganese and iron peroxides are distributed on the ocean floor as grains, nodules, slabs, coating-on-rocks, impregnations of porous materials, replacement fillings of coral and organic debris, and in other less important forms. Small manganese-dioxide grains, about 0.5 mm in diameter, are an almost invariable constituent of red clay and a common constituent of the organic oozes found in the pelagic areas of the world's oceans. Sea-floor rock outcrops are frequently coated with a layer of manganese and iron oxides sometimes 10–15 cm in thickness. Hard rocks found on the sea floor, such as boulders of granite or sandstone rafted to sea by icebergs or seaweed, are commonly coated with manganese oxides. Good conductors of electricity such as naval-shell fragments, which are sometimes dredged from the sea floor, can accumulate a coating of manganese and/or iron oxides several millimetres thick in the span of a few tens of years.

Fig. 38. Manganese nodules from the Atlantic Ocean. The location data are: A. N 30°, W 76°; depth 2,645 m. B. N 30°51′, W 78°27′; depth 732 m. C. N 29°17′, W 57°23′; depth 5,840 m. D. N 32°13′, W 69°06′; depth 5,290 m. E. N 20°24′, W 66°24′; depth 5,520 m. F. S 49°21′, W 47°45′; depth 4,840 m.

Pumice stones and blocks are frequently coated and impregnated with manganese–iron oxide minerals; so also is coral debris. From an economic standpoint, because of technical considerations in the mining system, nodules are the most important of the forms of the sea-floor manganese–iron oxides.

Physical characteristics of the nodules

Manganese nodules are found in a variety of physical forms. Agglomerating colloidal particles tend to form a spherically-shaped concretion. In the ocean, however, there are many influences operating to modify the spherical shape. The nodules have been described

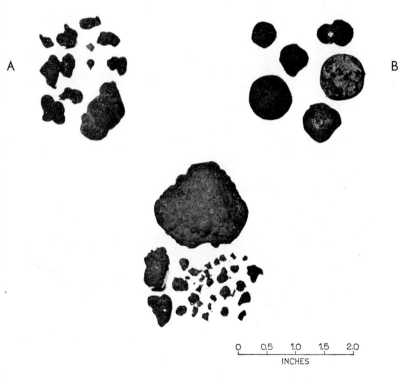

Fig. 39. Manganese nodules from the Indian Ocean. Location data are: A. S 26°54′, E 56°04′; depth 4,855 m. B. S 29°52′, E 62°36′; depth 4,396 m. C. S 37°50′, E 124°30′; depth 5,518 m.

Fig. 40. Manganese nodules from the North Pacific Ocean. Location data are: A. N 29°58′, W 125°55′; depth 4,325 m. B. N 23°17′, W 138°15′; depth 4,890 m. C. N 22°30′, W 113°08′; depth 3,600 m. D. N 14°11, W 161°08′; depth 5,652 m. E. N 9°57′, W 137°47′; depth 4,930 m. F. N 21°27′, W 126°43′; depth 4,300 m.

MANGANESE NODULES 131

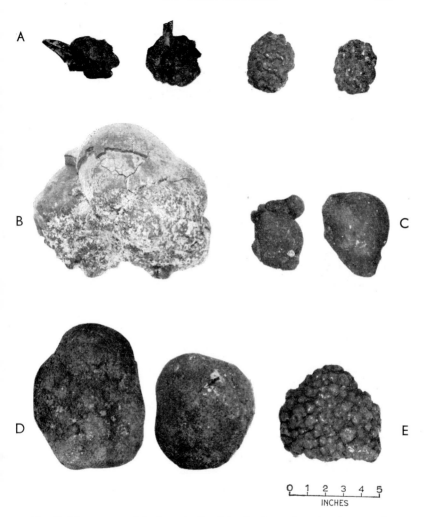

Fig. 41. Manganese nodules from the South Pacific Ocean. Location data are: A. S 16°29′, W 146°33′; depth 1,270 m. B. S 8°30′, W 85°36′; depth 4,330 m. C. S 9°00′, W 171°28′; depth 5,000 m. D. S 18°55′, W 146°23′; depth 4,460 m. E. S 41°59′, W 102°01′; depth 4,200 m.

as appearing like potatoes, mammillated cannon balls, marbles, tablets, and a number of other less recognizable forms. Although nodules from a particular locality often exhibit a group resemblance in size, shape, and appearance, nodules from different parts of the ocean tend to have unique physical characteristics. Fig. 38, 39, 40 and 41 illustrate a few of the forms in which the nodules appear.

If a nodule has a sizeable nucleus, the shape of the nucleus is followed by the growing nodule. Many of the nodules shown in Fig. 38–41, however, exhibited no discernible nucleus when they were studied in cross-section. Slab-like nodules are common in certain areas of the Pacific Ocean. This shape may be due to several causes, the most common one probably being the coalescence of a closely-spaced group of similar-diameter nodules. When an intermixture of uniform diameter nodules and slabs appear, as in Fig. 42, the slabs generally prove to be mangenese-dioxide encrusted blocks of pumice.

Manganese nodules from the sea floor are generally earthy black in color; however, their color may vary from black to tan. Nodules with a high-iron content generally are reddish-brown while nodules with a high-manganese content are blue–black. Most of the nodules are dull in appearance; however, nodules from an area about 400 km northeast of Tahiti and from the Blake Plateau off the east coast of the United States have a vitreous luster.

The hardness of the nodules is variable, ranging from one to about four on the Mohs scale. An average hardness would be about three. Nodules, high in calcium carbonate, which seems to act as a cementing agent when present in the nodules in amounts greater than about 5%, are relatively hard and difficult to crush. Nodules with a calcium carbonate content less than 2 or 3% generally are very friable and easily crushed.

In size, the nodules generally range between 0.5–25 cm in diameter, but average about 3 cm if only those nodules which have no nucleus or only a relatively small one are considered. This apparent size range may be somewhat influenced by the sampling devices that have been used to recover the nodules. No object much larger than 1 ft. in diameter can enter the 1 × 3 ft. mouth of the chain bag dredge and nodules less than 1 cm in diameter would tend to filter through the openings in the net of this dredge. Photographs have shown nodule-like objects on the sea floor which are 1.5–2 m in diameter,

Fig. 42. A sea-floor photo taken at N 19°48′, W 120°16′; depth 4,104 m, showing an intermixture of relatively uniform diameter nodules with irregularly shaped blocks. The blocks are very likely composed mainly of pumice which has floated to this location from a distant volcanic eruption before becoming water logged and sinking to the sea floor. Once on the sea floor, the pumice block can accumulate a coating of manganese–iron oxides. (Photo by Nikita Zenkevitch, Institute of Oceanology, Moscow, U.S.S.R.).

as shown in Fig. 43, but these objects could be merely manganese-dioxide encrusted boulders or rock outcrops.

The largest nodule ever recovered was a 850-kg mass which was found entangled in a telegraph cable being salvaged about 500 km east of the Philippine Islands. After having been sketched and sampled, this nodule was cast back into the sea (B. Heezen, personal communication, 1958). The largest known nodule ever recovered and kept was dredged from the Blake Plateau off the east coast of

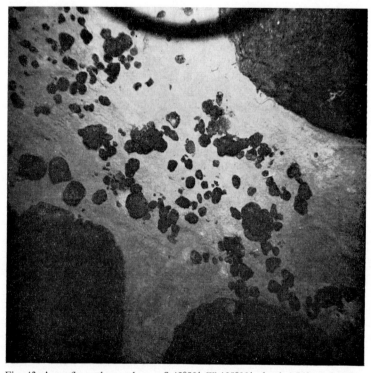

Fig. 43. A sea-floor photo taken at S 42°50′, W 125°32′; depth 4,560 m, showing manganese nodules and what is apparently manganese dioxide encrusted boulders or rock outcrops. The photograph covers an area of about 4 × 5 ft. and, if the object in the upper right hand corner is spherical, it must be at least 6 ft. in diameter. It is unlikely that it is a nodule that large. (Photo by Carl Shipek, U. S. Navy Electronics Laboratory, San Diego, California).

the United States. It weighed 55 kg and, on cross-sectioning, was found to be solid manganite. Another large nodule, called the Horizon nodule, is shown in Fig. 44. It was recovered in a tangle of core-barrel wire, an excess of which had been let out by mistake. It weighed about 45 kg but was found to have an indurated zeolite core of over half the volume of the nodule. It was recovered at N 40°14′, W 155°05′, about 2,400 km north of Hawaii. Radioactive elements such as radium and thorium are incorporated in the layers of the manganese nodules as they form. By assaying for these elements, GOLDBERG (1961b) was able to date various horizons in this nodule. He concluded that the Horizon nodule formed at an average rate of about 0.01 mm per 1,000 years. At this rate the nodule would require about 16 million years to reach its present size. GOLDBERG (1963b), however, indicated that the growth rate probably was not uniform and that the nodule may not have increased in size appreciably in the past 500,000 years.

Given proper environmental conditions, manganese nodules could probably accrete to diameters in the order of meters. Present information, however, would indicate that such nodules are unlikely to be found in any great numbers. The majority of the nodules thus far recovered by dredging do not exceed about 8 cm in diameter and the thickness of crusts torn from rock outcrops on the sea floor does not frequently exceed 10 cm. As viewed in sea-floor photographs, the nodules generally average between 2–4 cm in diameter.

The density of the group of nodules on which analyses are listed in Table XXX averages 2.49 g per cm^3, but ranges between 2.07–3.07. Nodules from a given deposit seem to have a uniform density. This observation, however, is based on tests performed on a group of nodules from only a single deposit.

Nuclei of the nodules

The chemical nature of the nucleus does not seem to affect the deposition of manganese–iron oxides or the composition of the nodule. The nuclei may be carbonates, phosphates, zeolites, clays, or various forms of silica. Any hard object seems to be able to serve as a nucleus. MURRAY and RENARD (1891) note that basic and acid

silicates, such as pumice and glassy lapilli, almost always profoundly altered, seem to be the most frequent nuclei. These nuclei are followed, in order of abundance, by the teeth of sharks and other fish, otoliths, bones of Cetaceans, and siliceous and calcareous sponges. Casts of Foraminifera have been recognized as nuclei in the nodules. Manganese-oxide deposits also collect on all forms of hard rock found on the sea floor, as well as on materials dropped into the sea from ships. In many nodules, no nucleus can be observed on cross-sectioning. Apparently the colloidal precipitates themselves or microscopic grains of sediments can also serve as nuclei.

In many cases, the external form of the nodule depends on the shape of the nucleus. Sometimes the nucleus is not a single body, but several. When these nuclei are near one another, the growing nodules frequently coalesce into a single nodule with several knobs. The surfaces of the nodules are generally covered by all sorts of spicules and mammillae.

Not only the presence of nuclei, but the external and internal appearance, especially the botryoidal and cryptocrystalline characteristics, indicate that the nodules are concretionary in nature. Cross-sections show the nodules are built up of successive concentric shells. This onion-skin structure, illustrated in Fig. 45, is common of materials in nature formed by the agglomeration of colloidal particles. This structure, however, is common also to manganese-dioxide deposits formed by the action of bacteria (KALINENKO, 1949). Interbedded in the shells of the nodules are layers of clay and ooze, which layers are generally less than 0.1 mm in thickness. When a nodule is crushed or broken, the shells tend to part along these interbedded layers.

The individual shells of manganese–iron oxides are generally not uniform in thickness, indicating a variable rate of formation of the nodules, discontinuous periods of growth, and/or great changes in

---

Fig. 44. Several views of the Horizon nodule. Fig. 44A shows a side view of the nodule, Fig. 44B a bottom view, and Fig. 44C a cross-sectional view. The white block in the interior of the nodule was mainly phillipsite. As shown in the cross-sectional view, the bottom layer of manganite is about 5 mm thick and the top layer, about 16 cm thick. The scale in the bottom photo is a 12-inch ruler.

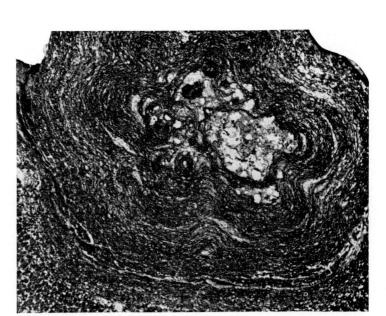

Fig. 45. Cross-sectional views of manganese nodules from N 20°24′, W 66°24′; depth 5,520 m, in the Atlantic Ocean just north of Puerto Rico. In many cases, the nodules seem to be composed of two distinct parts, an outer shell which is friable, and a more coherent inner core. The outer shell is partly broken away in the upper photo. The bottom photo shows the onion-skin growth structure common of all ocean-floor manganese nodules. The magnification of the bottom photo is 4 ×.

the rate of formation of the associated sediments. At high magnifications, as shown in Fig. 46, the individual shells appear to be oolitic, a structure common in continental wad deposits.

ENVIRONMENTAL FACTORS OF FORMATION OF MANGANESE NODULES

Manganese nodules of one form or another are found in all ocean sediments formed under oxidizing conditions. Manganese-dioxide grains are a common constituent of the terrigenous sediments near the continents. Terrigenous sediments, however, generally form at a much greater rate than do the nodules. Once a nodule is buried in sea-floor sediments, it will cease to enlarge by a process of particle

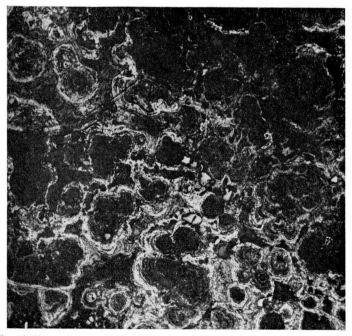

Fig. 46. Photomicrograph of a polished surface of part of the shell of the nodule shown in Fig. 45. The concentrically banded spherulites of manganese–iron oxides grow around minute grains of detrital materials; × 55.

agglomeration. Nodules, thus, will grow to appreciable sizes at those locations in which the rate of formation of the associated sediments is less than the rate of formation of the nodules, or where the nodules, by some external means, can remain at the surface of the associated sediments.

Associated sediments

Much of the data concerning manganese-nodule deposits can be found in the *"Challenger" Reports* (MURRAY and RENARD, 1891) and in the *"Albatross" Reports* (AGASSIZ, 1901, 1906; MURRAY and LEE, 1909). In these reports are presented tables and maps describing the sediments recovered at each station where dredging was done. From these tables can be drawn statistics which shed light on the distribution of manganese nodules with respect to sediment types in the ocean basins. Table XXI lists statistics concerning the occurrence of manganese nodules and grains in the Pacific pelagic sediments. Manganese grains would be noted more often than nodules, as at many stations only a sediment sample was taken. Also listed in Table XXI are statistics concerning the distribution of MnO_2 grains in the sediments of the Atlantic Ocean. Generally, manganese grains are associated with manganese nodules of larger size, but the reverse is not necessarily true.

By the data of Table XXI it appears that the sediment type does exert some influence over the formation of the nodules. This control is most likely due to the rate of formation of the sediment or the activity of the agency keeping the nodules at the surface of the sediment in a particular locality and not due to the processes forming the sediment or the chemical nature of the sediment. As would be expected, the incidence of manganese nodules and grains is greatest with the red-clay deposits. This association is most likely the result of the fact that the red clays are very slow, in the order of millimeters or even tenths of a millimeter per 1,000 years, in forming.

An unusually high incidence of manganese nodules and radiolarian ooze is also indicated, however, the small number of samples taken in this type of sediment may have biased this result.

The southeast Pacific-Ocean floor, an area largely covered by calcareous oozes, seems to be covered by deposits of manganese nodules. In this region, nodules generally are recovered at every dredge station and also at a great many of the core stations indicating very high concentrations of nodules. The rate of formation of the calcareous sediments is usually measured in terms of centimeters per 1,000 years; thus, the rate of formation of the nodules in these areas must be relatively rapid and the appearance of the nodules a compara-

TABLE XXI

DISTRIBUTION OF MnO_2 GRAINS AND MANGANESE NODULES IN THE PELAGIC SEDIMENTS WITH REPSECT TO SEDIMENT TYPES

Sediment type	Number of samples	With MnO_2 grains		With manganese nodules	
		number	percentage	number	percentage
"Challenger" Expedition Reports[1]					
Red clay	45	39	87	16	36
Globigerina ooze	23	15	65	6	26
Blue mud	13	7	54	2	15
Radiolarian ooze	7	6	86	6	86
Other[2]	6	0	0	0	0
"Albatross" Expedition Reports[3]					
Red clay	17	—	—	14	82
Globigerina ooze	34	—	—	15	44
Diatom ooze	14	—	—	5	28
Radiolarian ooze	3	—	—	3	100
Other[4]	12	—	—	0	0
In the Atlantic Ocean[5]					
Red clay	126	54	43	—	—
Globigerina ooze	772	43	6	—	—
Blue mud	342	12	4	—	—
Pteropod ooze	40	6	15	—	—
Volcanic mud	102	1	1	—	—
Other[6]	44	0	0	—	—

[1] Abstracted from MURRAY and RENARD (1891). Pacific Ocean data only.
[2] Largely volcanic muds. Only sediment samples from depths greater than 1,830 m were included.
[3] Abstracted from AGASSIZ (1906). All samples from southeast Pacific.
[4] Volcanic mud, rocky ground, green mud, blue mud.
[5] Abstracted from DIETZ (1955).
[6] Green mud, coral mud, diatom ooze, red mud, calcareous sand.

142   THE DEEP-SEA FLOOR

tively recent phenomenon in geologic history, or the nodules must have some method of rising as the associated sediment is formed to remain at the surface of the sea floor.

Ocean-floor currents

In many of the sea-floor photographs, scour marks and ripples in the fine sediments around the nodules indicate the presence of appreciable water currents; such scour marks can be clearly seen in Fig. 47. Water currents can serve in several ways to promote the formation of manganese nodules; they sweep away extraneous sediments but allow the charged manganese and iron-oxide particles to accrete on attractive surfaces, they carry a fresh supply of manganese and iron

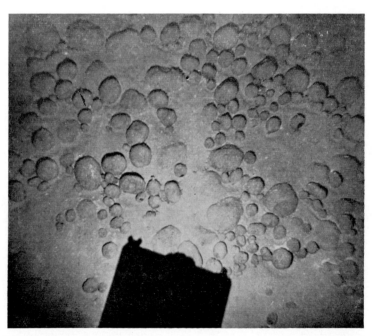

Fig. 47. Manganese nodules on the Atlantic Ocean floor at N 30°37′, W 59°07′, depth 5,490 m, about 300 miles southeast of Bermuda. Scour marks in the sediments around these nodules indicate an appreciable bottom water current. The area covered by this photograph is about 4 × 5 ft. (Photo by David Owen, Woods Hole Oceanographic Institution, Woods Hole, Massachusetts).

ENVIRONMENTAL FACTORS OF NODULE FORMATION

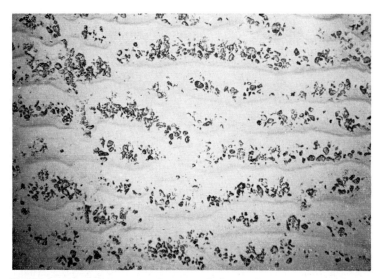

Fig. 48. A sea-floor photograph taken at N 10°02′, W 165°29′, depth 1,640 m, showing waves in calcareous sediments. The dark objects, apparently concentrated in the troughs of the waves, are manganese nodules. Most likely, the nodules are as concentrated under the crests of the sediment ripples.

to the nodule-forming area, and they help to maintain an oxidizing atmosphere on the ocean floor. If a reducing atmosphere were set up by stagnant waters, the manganese nodules would likely dissolve.

Ocean-floor currents can also retard growth of the nodules by sweeping sediments over the nodules, thus isolating them from sea water. Ripples in sediments of the ocean floor frequently are found in the vicinity of manganese nodules. Fig. 48 shows manganese nodules apparently concentrated in the trough of sediment waves; the bed of nodules, however, is probably as concentrated under the crests of these waves.

The concentric, layered structure of the nodules is indication of a succession of periods of growth separated by intervals of quiescence (CHOW and PATTERSON, 1959; GOLDBERG and KOIDE, 1958). Such processes could be explained by sediment waves moving slowly over the ocean floor alternately covering and exposing the nodules. While the nodules are under the crest of the sediment wave, they will cease to grow. As the wave moves on and the nodules are uncovered in the trough, the growth process can continue.

SKORNYAKOVA (1960) noted the marked association of nodules with the areas of abyssal hills. She remarked that this association was probably due to the fact that the dissected nature of the hills requires a very low rate of sediment accumulation and that volcanic rocks, which could serve as a source of manganese for the nodules, frequently crop out in these regions.

Effects of animals on the ocean floor

Another common environmental characteristic of the sediments in the vicinity of many nodule deposits is the evidence of animal activity. Many investigators have described a large number of benthic organisms which plow, burrow, and eat their way through deep-sea sediments, thereby extensively reworking them (BRAMLETTE, 1961; BRUUN and WOLFF, 1961; SOKOLOVA, 1959; SPÄRCK, 1956; ZENKEVITCH, 1961). ARRHENIUS (1963) describes sediment cores in which burrowing animals have almost completely destroyed the layered structure of the sediment. Fig. 49 illustrates a common surficial result of animal activity in the sea-floor sediments.

A number of these sea-floor animals are mud eaters which ingest the surficial sediments, extract the organic matter of food value and expel the remainder. These animals are of appreciable size and it would be expected that if nodules were encountered as the animals made their way through the sediments, the nodules would be pushed aside or ingested and expelled along with the other non-organic constituents of the mud. In either case, the nodule should come to rest in a position somewhat different than when encountered by the animal. Also, the nodule would experience a net upward movement to the surface of the sediments.

A number of organisms, such as worms, live within the top 25 cm or so of the pelagic sediments. In burrowing through these sediments, the action of the animal, if it comes in contact with the bottom portion of a nodule, would be to push the nodule upward. The soft surrounding sediments would fill in under the nodule leaving it with a net upward movement again. The bearing strength of the pelagic

FORMATION OF MANGANESE NODULES 145

Fig. 49. A photograph taken at S 9°00′, E 171°28′, depth 5,000 m, showing mounds, holes, tracks, fecal pellets, and other evidence of extensive animal activity in these sediments. While nodules are not clearly visible in this photograph, about 25 of them were recovered at this station in a small dredge bucket which was tied to a leg of the camera with which this photo was taken. Apparently, a rather high concentration of nodules is covered by a thin layer of sediments at this location.

sediments is in the order of 50 g per cm² which is of sufficient magnitude to support the nodules. There should be no tendency for the nodules to settle into the sediments. The only apparent agency working to bury the nodules is the continuous rain of sediment falling around and on them. Sometimes, as shown in Fig. 55, the nodules have a cap of sediment on their upper surfaces.

It is astonishing how the nodules in a deposit, as illustrated in Fig. 50, can be so regularly spaced and can grow to large diameters without coalescing and forming a solid crust on the sea floor. A process involving periodic disturbance of the growing nodules would explain this phenomenon. As the nodules grow at the rate of 1–0.001 mm per 1,000 years, the degree of animal activity would not need to be very great to keep the nodules properly disturbed and at the surface of the sediments.

FORMATION OF MANGANESE NODULES

The dominant agencies by which manganese is added to the ocean

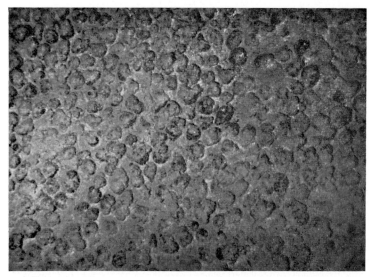

Fig. 50. A remarkable mosaic of manganese nodules at S 61°40′, E 170°40′; depth 5,160 m, about 1,000 miles south of New Zealand. The nodules range 2–8 cm in diameter and the concentration estimate is 2.5 g/cm². Although sediments are beginning to collect on the upper surfaces of these nodules, scour marks can also be seen around the edges of the nodules. (Photo by S. Calvert, Scripps Institution of Oceanography, U. S. Navy Photo).

are streams, submarine volcanic eruptions, submarine springs, and the decomposition of sea-floor igneous outcrops and debris. Sir John Murray thought that the submarine decomposition of manganese-rich basic eruptives was probably the most important source of the manganese in the nodules. Although these eruptives might be covered with sediment, MURRAY and RENARD (1891) proposed that manganese ions could diffuse to the sediment surface through the water in the pore spaces of the sediments. GOLDBERG and ARRHENIUS (1958) discount this idea and indicate that sufficient manganese and iron is added to the ocean by rivers and streams to explain that found in the sea water and in the sea-floor sediments. It is most probable, however, that a combination of the four processes listed is responsible for the manganese in sea water. Within certain localities, it is likely that one type of process predominates in furnishing manganese or iron for agglomeration in the nodules. For example, sea-floor springs off the southeast coast of Japan, in the Fuji volcanic zone, empty manganese-rich solutions into the ocean (NIINO, 1959). The high density of sea-floor manganese nodules and crusts in about 100 m of water in this locality is probably a result of the rapid precipitation of manganese from these spring waters as they enter the oxidizing atmosphere of the ocean. It is currently accepted by oceanographers, that the dissolution of igneous rocks, both on land and on the sea floor, is the major source of the manganese and iron in the sea. There are a number of differences, however, in the theories covering the transfer of the manganese and iron from the time they enter solution to the time they are fixed in an insoluble form in the manganese nodules. Few of these theories are based on experimental evidence. Prior to 1958, all were more or less deductions from observations made on the physical structure and bulk-chemical composition of the manganese nodules themselves or on similarly appearing material from the continents. Some experimental evidence was developed and presented (MURRAY and IRVINE, 1894; H. PETTERSSON, 1945), but, in the main, this evidence was used to disprove previously developed theories.

To explain the formation of the sea-floor manganese nodules, MURRAY and RENARD (1891) suggested that carbonate and sulphate bearing waters would act on the sea-floor rocks to release manganese, mainly as the bicarbonate. In the presence of the dissolved oxygen in the sea, the bicarbonate would be transformed to a colloidal manganese-peroxide which, in suspension in sea water, would become hydrated and would then collect on any convenient hard surface to which it was attracted. This theory appeals to geochemists (CLARKE, 1924; RANKAMA and SAHAMA, 1949) as it approximates the common geochemical cycle of manganese on the continents.

CORRENS (1941) suggested the possibility of a biological extraction of manganese from solution in sea water involving the ingestion of this element by Foraminifera and retention in the shells of these planktonic animals. Upon settling on the sea floor, the manganese-rich tests would redissolve with the manganous ion eventually being oxidized to an insoluble tetravalent state and collecting on available nuclei. Some evidence in support of a biological extraction of the metals from sea water has recently been presented by GRAHAM (1959), GRAHAM and COOPER (1959), and EHRLICH (1963). In support of this idea, several authors have noted that iron and manganese sols are stabilized by organic matter while in suspension in water (GRUNER, 1922). ASCHAN (1932) noted the importance of humic complexes in keeping manganese in solution in rivers. By feeding on the organic part of these complexes and oxidizing the manganese to the tetravalent state, bacterial agencies could cause the precipitation of manganese from solution. This process is apparently the one followed by bacteria in the precipitation of manganese from fresh water solutions on the continents.

Other processes of note that have been proposed are: the deposition of manganous sulfide with subsequent oxidation (BUCHANAN, 1890), bacterial oxidation (BUTKEVITSCH, 1928; DIEULAFAIT, 1883; DORFF, 1935), and inorganic oxidation and precipitation (BOUSSINGAULT, 1882; KUENEN, 1950; MURRAY and RENARD, 1891). CASPARI (1910) suggested that the nodules might be the result of the dissolution of the soluble silica leaving volcanic debris enriched in manganese, iron and other metal oxides. H. PETTERSSON (1945), critically examined all of the known theories of the formation of sea-floor manganese and iron oxides and rejected most of them. He then proposed that small particles of volcanic material settling through the water could act as centers of accretion and would be effective in extracting manganese and iron sols from the sea water. That such small particles of volcanic ejecta are spread over large portions of the world at the time of large volcanic eruptions is well known (MENARD, 1960; RICHARDS, 1958). Fine dust from the Krakatoa eruption, in 1883, was known to have settled over most of the surface of the earth. Wind-borne dust from erosion on the continents is also distributed over large areas of the ocean surface (ARRHENIUS, 1959).

A major difficulty of these hypotheses is that they were formulated without adequate knowledge of either the mineralogy of the nodules or the chemical nature in sea water of the elements involved. Many of the hypotheses do not take into account the presence in the nodules of elements other than iron and manganese. Fundamental studies of the mineralogical composition of the nodules and the nature of the agglomerating processes were started in the 1950's (BUSER and GRÜTTER, 1956; GOLDBERG, 1954; GRÜTTER and BUSER, 1957; KRAUSKOPF, 1956) and recently, more detailed theories accounting for the formation of the nodules have been developed.

To explain the presence of metals in addition to manganese and iron in the nodules, GOLDBERG (1954) developed a theory based on data concerning chemical scavengers. The action of certain colloidal particles in removing specific elements from solution is well known to chemists (COOK and DUNCAN, 1952), while the great efficiency of certain scavenging agents in removing specific ionic species from solution in sea water has been demonstrated by KRAUSKOPF (1956).

At its slightly alkaline pH of about 8, and with its highly oxidizing atmosphere as the result of the dissolved oxygen, sea water is saturated with iron and manganese in solution. The concentrating effects of evaporation force the precipitation of manganese and iron. In precipitating, these elements apparently form colloidal particles. The presence in sea water of particulate iron hydroxides had been demonstrated by COOPER (1948) and by GOLDBERG (1952), while the presence in sea water of particulate, hydrated manganese oxides has been demonstrated by RANKAMA and SAHAMA (1950) and GOLDBERG and ARRHENIUS (1958). These colloidal particles are electrically charged and, while filtering down through the sea water, tend to act as scavengers and remove certain ionic species such as copper, nickel, cobalt, molybdenum, zinc, lead, etc., from solution in the sea water. These elements are vastly undersaturated in sea water (GOLDBERG, 1954; KRAUSKOPF, 1956), consequently, some process such as scavenging or bacterial extraction must be invoked to get them into and hold them in the manganese nodules.

Once on the ocean floor, the colloids of iron and manganese are swept along by the bottom-water currents until they come into contact with some surface capable of holding them. GOLDBERG (1954) suggests that hard objects protruding above the ocean floor would act as superior electric conductors in attracting the electrically charged colloidal particles. It is a matter of observation, that metallic objects, which would be superior electrical conductors, accrete manganese and iron oxides at a much greater rate than do non-metallic nuclei. It may be that these hard surfaces possess chemical properties which make them favorable localities for the agglomeration of iron and manganese. Also, bacteria tend to congregate on hard-solid objects where they make their living by extracting organic matter from the sea water and by oxidizing oxidizable material such as the manganous ion. If bacteria do play an important role in the formation of the nodules, and at present there is insufficient direct evidence to assume that they do, the hard objects would be preferred as nuclei simply because they are centers of bacterial populations and not because of any special chemical or electrical properties that they might possess. If it can be shown that bacteria do play a dominant role in the formation of the nodules, it would explain many observations concerning the nodules such as their concentric-shell structure, the relatively uniform shell thickness, and the appearance of the divalent metal ions commonly attracted by chelating agents which bacteria are known to produce (JONES, personal communication, 1962).

In view of the more recent data of geochemistry and in the light of work by BUSER AND GRÜTTER (1956) on the mineralogy of the nodules, GOLDBERG and ARRHENIUS (1958) have constructed a much more complex theory on the method of manganese-ion transfer. Experiments by Goldberg and Arrhenius indicate that over 85% of the manganese in sea water is in the form of a true solution, presumably as a divalent ion; the other 15% apparently prevaling in sea water in a colloidal form. BUSER and GRÜTTER (1956) describe the mineralogy of the nodules as a double-layer structure where layers of MnO_2 alternate with disordered layers of hydrated $Mn(OH)_2$ and hydrated $Fe(OH)_3$. The major part of the manganese in the nodules is, therefore, in a tetravalent form. That tetravalent manganese is not found in sea water in considerable amounts but accretes at the sediment-water interface suggests the necessity of a reactive and oxidizing surface. The well-known catalytic properties of iron oxides, coupled with the frequent presence in the nodules of large amounts of iron, suggests that ferric oxides may provide the most important initial, active surface. Assuming that the oxidation requires catalysis on the oxide surface, Goldberg and Arrhenius suggest that the amount of manganese deposited will be related to the length of exposure of a given surface to sea water, and to the rate of supply of ferric oxides. Thus, if the iron accumulates at a high rate, the active surfaces will be continually replaced and high rates of manganese accumulation will be expected. In areas with low total rates of deposition and high contents of dissolved oxygen in the bottom waters, oxidation of dissolved manganous ions at a suitable surface and subsequent accretion of ferro-manganese oxide minerals may be the dominant accumulation process (GOLDBERG and ARRHENIUS, 1958).

Polyvalent ions with high charge densities such as copper, the rare earths, zinc, lead, etc., more readily enter into sorption reactions than the alkali or alkaline earth metals (GOLDSCHMIDT, 1954), and GOLDBERG and ARRHENIUS (1958) suggest that the forming $\delta\text{-MnO}_2$ acts as a scavenger of these high-charge-density ions.

The most detailed discussion of the form of various elements in sea water and the agencies which might transfer them from solution in sea water into the nodules is given by KRAUSKOPF (1956). He discussed the factors controlling the concentration of some thirteen metals in sea water, among which were Zn, Cu, Pb, Ni, Co, and Mo. His principal conclusion was that all of the elements he investigated were greatly undersaturated in sea water. In this same paper, Krauskopf describes a number of experiments in sea water with various scavengers and differing concentrations of the metals he investigated. He shows that copper, zinc, and lead are adsorbed more strongly and more consistently than are the other metals. All the adsorbents he tried, except plankton, reduce the concentrations of these three metals within a few hours nearly to the range found in sea water. Cobalt and nickel, although divalent ions, are much less effectively adsorbed than the other metals tested. For these two metals, however, manganese dioxide seemed to be, by far, the best adsorbent. Considering the scavengers tested, manganese dioxide was the best adsorbent followed by iron oxide.

MINERALOGY OF MANGANESE NODULES

MURRAY and RENARD (1891) classified the manganese nodules among the impure varieties of manganese oxides known as wad or bog manganese ore. The name, wad, is generally applied to manganese deposits occurring as amorphous and reniform masses. Wads are largely mixtures of manganese peroxides, limonite, detrital materials, and often include small percentages of cobalt, nickel, and copper. They are related to psilomelane but are mixtures of different oxides and cannot be considered distinct mineral species.

Later work, principally by BUSER (1959), BUSER and GRÜTTER (1956), GOLDBERG and ARRHENIUS (1958), GRÜTTER and BUSER (1957), and RILEY and SINHASENI (1958), has shown the nodules to consist of a number of minerals and the nodules are therefore most appropriately classified as rock. Besides detrital minerals of silica and alumina such minerals as opal, goethite, rutile, anatase, barite, nontronite, and at least three manganese-oxide minerals of major importance, have been recognized in the nodules (ARRHENIUS, 1963). These minerals are generally found in the nodules as intimately intergrown crystallites.

Although the very fine-grained structure of the nodules makes mineral identification by X-ray analysis difficult, techniques, recently developed, have been used to show that two of the three major manganese minerals are not similar to any of the presently known minerals of this element (GOLDBERG, 1954). It is indicated that these manganese minerals are a new and unnamed mineral species. BUSER and GRÜTTER (1956) and BUSER (1959) describe these minerals as having a layer-lattice structure with alternating sheets of ordered manganese dioxide and disordered iron-manganese oxides. The main layers, which contain Mn^{4+} in six-fold co-ordination with O^{2-}, are separated by 10 Å and contain between them a disordered layer of Mn^{2+} coordinated with water, hydroxyl, and possibly other anions (BUSER and GRÜTTER, 1956). This structure is similar to that of the mineral lithiorphoiite, and to certain synthetic minerals known as manganites. Various elements such as Na, Ca, Sr, Cu, Cd, Co, Ni, and Mo can apparently substitute for Mn or Fe in the disordered phase of the structure. Due to the disorder of these minerals it has not been possible to describe their structure with any degree of accuracy thus far (ARRHENIUS, 1963). When the nodules contain more iron than is needed to satisfy the manganite phase of the nodules, the mineral goethite appears. In some of the nodules a considerable quantitity of goethite is present (BUSER, 1959).

The third manganese mineral in the nodules seems to be identical with a synthetic manganese mineral described as δ-MnO_2 and is an aggregate of randomly oriented sheet units with an average size of about 50–100 Å (ARRHENIUS, 1963). In nodules with a small amount of iron, δ-MnO_2 seems to be the dominant mineral.

The cryptocrystalline nature of the oxide constituents and the similarity of the optical data of different manganese mineral species make optical identification of the various mineral species an impossibility. Opaque constituents form more than 90% of the volume of the shell portion of the nodules. Embedded in the opaque-manganite matrix are fragments of detrital minerals. The size of the detrital constituents ranges from submicroscopic to half a millimeter in diameter. The average size is in the order of 0.05–0.1 mm. X-ray diffraction patterns of nodule material are generally not revealing of any overall crystalline structure. While broad lines of manganese minerals can be recognized, only the detrital fractions of the nodules yield distinct lines.

The hydrochloric-acid-insoluble fraction of the nodules, which averages about 25% of the bulk weight of the nodules, is practically free of the heavy metals that are characteristic of the acid-soluble fraction. The HCl-insoluble fraction consists principally of clay minerals together with lesser amounts of quartz, apatite, biotite, pyroxene, hornblende, mica, spinels, rutile, anatase, and sodium and potassium feldspars (ARRHENIUS, 1963; RILEY and SINHASENI, 1958). The mineral assemblage composing the HCl-insoluble fraction of the nodules are largely detrital materials which are swept into contact with the nodules during their growth process and are incorporated in the physical structure of the nodules. A few of these minerals, such as rutile, however, may be authigenic.

Rate of formation of the nodules

One of the elements scavenged from sea water by the precipitating iron and manganese colloids is ^{226}Ra which is radioactive with a half-life of 1,600 years. By assaying various shells within the nodules for this element and by computing the ratio of the amount of ^{226}Ra present in the various shells to that of the amount present in the outermost layer, it is possible to determine the age of the shells. The accuracy of this method depends on a number of assumptions being true as: (*1*) the rate of accumulation of radium has been constant over the time span involved; (*2*) the growth process of the nodule has been continuous; (*3*) ^{226}Ra is unsupported by its parent, ionium, in the nodules; and (*4*) diagenetic chemical processes have not affected the concentration of radium in the various shells of the nodules. Using this method, H. PETTERSSON (1943) determined a rate of growth for one nodule of about 1 mm per 1,000 years. VON BUTTLAR and HOUTERMANS (1950) confirmed Pettersson's measurement, obtaining similar results (0.6–1.3 mm per 1,000 years) on several other nodules.

GOLDBERG and PICCIOTTO (1955) point out that the sols forming the nodules also scavenge thorium from sea water and suggest that the radium in the nodules is supported by ionium, the parent of ^{226}Ra, which is also radioactive but with a half-life of 80,000 years. Using this assumption, GOLDBERG (1961b) has determined rates of formation of the nodules as small as 1 mm per 100,000 years.

At the other extreme, a fragment of a naval shell was recovered from the sea floor, which had a coating of manganese–iron oxides almost 3 cm thick. As the type of shell that the fragment was part of was in use about 50 years ago, a rate of formation of the nodule of several centimeters per 100 years is indicated (GOLDBERG and ARRHENIUS, 1958). Another shell of the type used during World War II was found in about 200 m of water off San Diego, which had a coating of iron-manganese oxides about 1.5 cm thick indicating a rate of growth of almost 10 cm per 100 years. This shell was found on Forty Mile Bank where a highly oxidizing atmosphere exists and where water currents are generally measured in terms of a few knots.

The conclusion seems to be that there is no fixed rate of formation of the nodules for the ocean as a whole. The rate of formation is apparently a function of the quantity of manganese and iron available for agglomeration per unit of time and the electrical, chemical, and, possibly, biological properties of the surface attracting the manganese and iron-oxide particles. As the environment, in regard to these factors, will differ from place to place in the ocean, we can expect that the rate of formation of the nodules will also vary from place to place.

DISTRIBUTION AND CONCENTRATION OF MANGANESE NODULES

Several methods can be used to obtain an estimate of the concentration of manganese nodules on the ocean floor. Dredge and trawl hauls normally can only give rough indications of heavy or light concentrations of the nodules within an area. This author has obtained drag dredge hauls ranging from a few pounds to more than a ton of nodules, using the same bucket and dredging in the same area, which gives some indication of the lack of reliability of this method as a concentration-measuring device. In drag dredging, the bucket is lowered to the sea floor and dragged over the ocean floor for some distance. It is almost impossible to determine over what area of the sea floor the dredge bucket was in contact with the bottom while it was being towed. It is generally assumed, that the bucket will be riding above the sea floor at least part of the time.

Photography is commonly used as a means to measure the concentration of manganese nodules on the sea floor. Measurements from sea-floor photographs can be expected to yield an estimate within plus or minus 50% of the true concentration. In sea-floor photographs it can be assumed that not all the nodules will be discernible to the eye, especially the smaller ones. Scales are often confusing and rarely exact. Frequently the photograph is taken with the plane of the camera lens at an angle to the sea floor, which complicates the scaling problem. The shape of the nodules is generally irregular enough to make volume calculations from photographic gathered data somewhat speculative. Unless a sample of the nodules is available from the area in which the photograph is made, the specific gravity of the nodules cannot be determined. If the nodules are piled up in depth, as they might be in some of the abyssal-hill regions, the sea-floor photo would not likely disclose this fact.

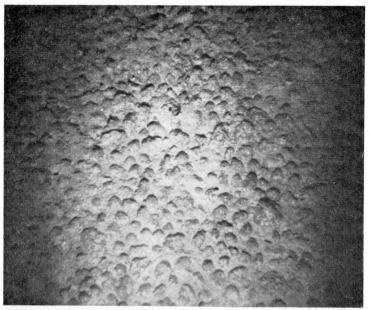

Fig. 51. Legend see next page.

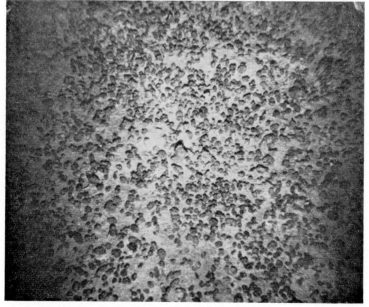

Fig. 52. Legend see next page.

Fig. 53. This photograph shows an oblique view of the ocean floor at N 10°25′, W 130°35′; depth 4,712 m. The nodules are about 3–10 cm in diameter. The nodules are resting on a clayey radiolarian ooze in the Clipperton Fracture Zone of the Pacific. (Photo by Carl Shipek, U. S. Navy Electronics Laboratory, San Diego, California).

Fig. 51. The area covered by this photograph, taken at N 20°00′, W 113°57′; depth 3,778 m, is 1.6 × 1.6 m or about 28 square ft. The concentration estimate is 1 g/cm². (Photo by Nikita Zenkevitch, Institute of Oceanology, Moscow, U.S.S.R.).

Fig. 52. Manganese nodules on the sea floor at N 19°57′, W 126°06′, depth 4,545 m. The area covered by the photograph is 1.6 × 1.6 m and the concentration estimate is about 1 g/cm². The angular shape of the nodules indicates a good deal of pumice at this point on the sea floor. (Photo by Nikita Zenkevitch, Institute of Oceanology, Moscow, U.S.S.R.).

158   THE DEEP-SEA FLOOR

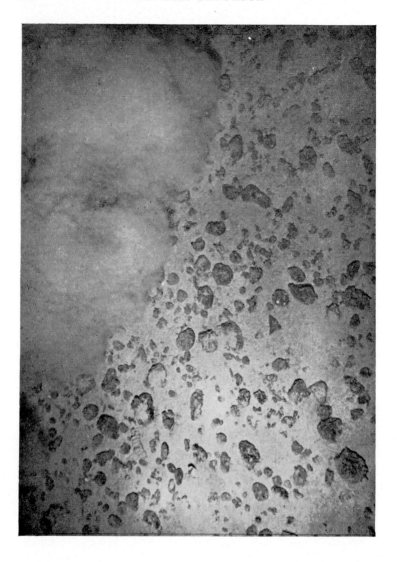

Fig. 54. The manganese nodules shown in this photograph are 1–10 cm in diameter. A shark tooth is visible in the center part of the photo. The sediment cloud was raised when the camera struck bottom. This photo was taken at N 20°38′, W 130°46′; depth 5,180 m. (Photo by S. Calvert, Scripps Institution of Oceanography, U. S. Navy photo).

NODULE DISTRIBUTION AND CONCENTRATION 159

Fig. 55. A sea-floor photograph taken at S 21°37′, W 147°40′; depth 4,684 m, showing manganese nodules with diameters of 2–10 cm. The nodules seem to be about half buried in the sediment and most have a capping of red clay. (Photo by Carl Shipek, U. S. Navy Electronics Laboratory, San Diego, California).

In favor of the camera is the comparatively large sampling area, 10–150 square feet per photograph as compared with 0.1–3 square feet for clamshell samplers. Considering that as many as 1,000 photographs can be taken on a single lowering of the camera, the sampled area per lowering can be of an order of 100,000 square feet or about 10^5 times as great as with the sediment sampling devices. Photographs also provide a wealth of information concerning sea-floor environmental conditions, sediment types, and water currents.

A number of investigators have developed deep-sea cameras

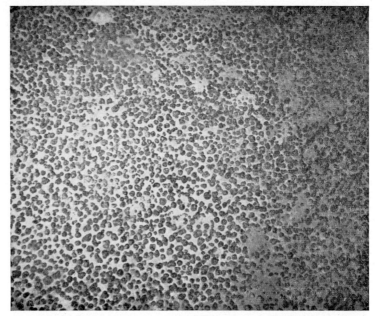

Fig. 56. This photograph shows a remarkably uniform distribution of nodules 2–5 cm in diameter. The concentration estimate is 1.5 g/cm². The white mounds covering the nodules in a few places are sediments probably thrown up by some burrowing animal. The photo was taken at S 13°53′, W 150°35′; depth 3,695 m. (Photo by S. Calvert, Scripps Institution of Oceanography, U. S. Navy photo).

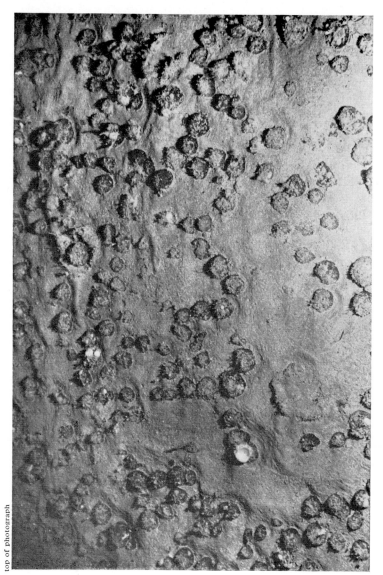

Fig. 58. This photograph was taken at N 31°03', W 78° 23'; depth 783 m, about 100 miles east of Jacksonville, Florida, on the Blake Plateau. Scour marks can be seen in the fine-grained calcareous sediments on which the nodules are resting. The nodules are 2–6 cm in diameter and the concentration estimate is about 2 g/cm².

← Fig. 57. A sea-floor photograph taken within about 100 ft. of the photograph shown in Fig. 56. Dredging, coring, and photography in this area of the ocean indicate a deposit of nodules covering at least several thousand square kilometers The larger scale of this photograph is due to a greater camera-to-sea-floor distance than for the photo shown in Fig. 56. (Photo by S. Calvert, Scripps Institution of Oceanography, U. S. Navy photo).

Fig. 59. This sea-floor photograph shows manganese nodules at N 29°17′, W 57°22′; depth 5,840 m, about 500 miles southeast of Bermuda. Scour marks can be seen in the sediments around the nodules, indicating appreciable water current velocities even at this great depth. The concentration estimate of the area (24 square ft.) shown in this photo is 1.4 g/cm². About 60 photos were taken at this station and they showed concentrations of the nodules ranging from 0.01 g/cm² to about 1.5 g/cm². The concentrations of nodules varied from station to station, almost as if the nodules were gathered in windrows. Sediment waves may be responsible for this phenomenon. (Photo by Dr. Bruce Heezen, Lamont Geological Observatory, Palisades, New York).

(EDGERTON, 1955; EWING et al., 1946; SHIPEK, 1960). In 1947, David Owen, of the Woods Hole Oceanographic Institute, took a remarkable photograph of the Atlantic Ocean floor in 5,500 m of water about 530 km southeast of Bermuda. That photograph, showing a concentration of nodules of about 5 g per cm² of sea floor, is reproduced in Fig. 47. An amount of 1 g per cm² is equivalent to 2.05 lb. per square foot or to 29,000 short tons per square mile. Since 1947, a number of investigators have photographed nodules on the sea floor in a variety of environments (DIETZ, 1955; ELMENDORF and HEEZEN, 1957; HAMILTON, 1956; HEEZEN et al., 1959; MENARD and SHIPEK,

1958; MERO, 1962; SHIPEK, 1960; ZENKEVITCH and SKORNYAKOVA, 1961). A sampling of the photographs obtained is shown in Fig. 50–60.

If the scale of a photograph can be determined, an estimate of the volume of manganese nodules on the sea floor, within the area shown by the photograph, can be made. An ellipsoidal shape is generally assumed for the nodules. Stereo pairs sometimes allow estimates to be made of the third dimension of non-spherical nodules. If a sample of the nodules is available from the area where the photograph was taken, the total weight of the nodules visible in the photograph can be calculated and a concentration of nodules per unit area of sea floor

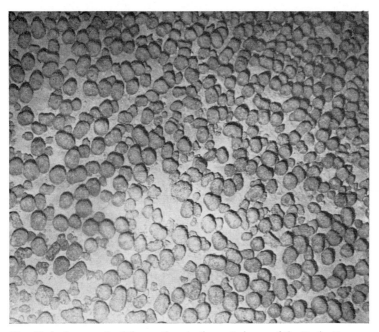

Fig. 60. A photograph of the sea floor in the central area of the Drake Passage, showing a remarkable deposit of manganese nodules lying on an apparently hard-packed bottom. The nodules stand out as if they had been eroded out of a sediment layer. The nodules are 1–5 cm in diameter and the concentration estimate is 1.5 g/cm^2. This photo was taken at S 57°59′, W 70°44′; depth 3,860 m. (Photo courtesy of the Chief Scientist, U. S. Antarctic Research Program, National Science Foundation).

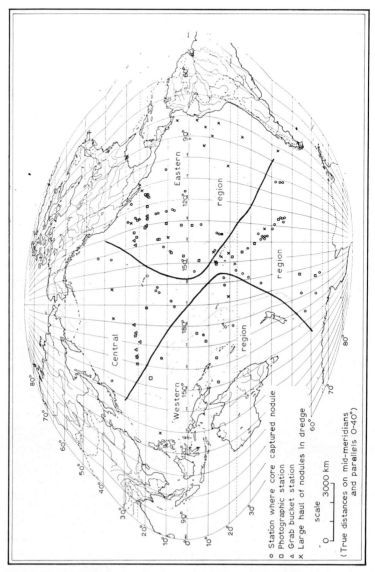

Fig. 61. Map of the Pacific Ocean, showing the locations of various sample points from which data were gathered concerning the concentration of manganese nodules at the surface of the sea-floor sediments. Also shown in this map are the concentrational regions of the manganese nodules in the Pacific.

can be determined. MERO (1960a) checked this method of estimating sea-floor concentrations of phosphorite concretions in the laboratory by laying nodules on a grid and photographing them. From the photographs, estimates were made of the areal concentration of the nodules. These estimates were compared with the actual concentration of the nodules on the grid. The estimated values varied from the true values by not over about 10% of the actual concentration indicating that reasonably accurate estimates of sea-floor concentrations can be made from photographs if the scale of the photograph is known and if all the nodules in the area covered by the photograph can be seen.

Surface concentration measurements on manganese nodules

Fig. 61 shows the location of available photographs taken of the sea floor in the Pacific Ocean. Table XXII lists the stations where surface concentration estimates have been made from photographs.

In addition to photography and clamshell sampling, sea-floor concentration estimates on the nodules can be made by an analysis of the number and size of the nodules captured in sediment cores taken within an area. Gravity cores, about 1 m long, are the most common type of sample taken of pelagic sediments. A relatively large store of cores can be found at oceanographic museums around the world. Frequently, nodules are found at the top of the sediment column in these cores. These cores are generally about 5 cm in diameter. Assuming that a direct hit on the single nodule within an area is an unlikely event, it requires a highly-concentrated deposit of nodules to allow capture by the corer. On the basis of statistical probabilities, the least concentration of spherically-shaped nodules averaging 2.0 cm in diameter which would present to a 5-cm diameter core barrel a 50% chance of capture would be about 0.22 g of nodules per cm^2 of sea floor, assuming a specific gravity of 2.4 for the nodules. The maximum concentration possible with a monolayer of 2.0 cm diameter nodules of specific gravity 2.4 is about 2.9 g per cm^2. By assuming a monolayer of evenly distributed, spherically-shaped nodules at the sediment-water interface and by calculating the concentrations for close-packed beds of nodules for the maximum concentrations and the spacing on the intersections of a 5-cm grid as the minimum

TABLE XXII

SEA-FLOOR CONCENTRATIONS OF MANGANESE NODULES AS DETERMINED BY BOTTOM PHOTOGRAPHY

Station[1]	Location data		Depth (m)	Average diameter of nodules (cm)	Area covered by nodules (%)	Concentration estimate (g/cm^2)	Remarks, references
	Latitude	Longitude					
Vit. 3632	N 17°38′	E 153°54′	5,718	—	26	0.46	
Vit. 4359	N 24°01′	E 162°02′	5,573	—	52	1.90	
Vit. 4347	N 24°01′	E 174°59′	5,383	1–5	38	0.92	
Vit. 4362	N 24°04′	E 160°46′	13,950	—	80	2.2	Skornyakova and Zenkevitch (1961)
Vit. 4331	N 19°57′	W 171°39′	3,680	—	50	0.90	
Vit. 4249	N 24°55′	W 132°18′	4,975	—	25	0.95	
Vit. 4285	N 19°57′	W 126°06′	4,545	—	21	0.36	
Vit. 4279	N 19°40′	W 120°16′	4,104	—	38	1.00	
Vit. 4273	N 20°00′	W 113°57′	3,780	2–3	34	0.60	
Expl. 14b	N 19°46′	W 114°44′	3,440	1–5	10–40	1.0	Nodules and rock
Expl. 14c	N 19°33′	W 114°28′	3,580	7	25	0.8	Uniform distribution
Expl. 14d	N 19°20′	W 114°12′	3,480	1–150	20–60	1.2	Nodules and rock
Msn 8-31	N 24°18′	W 126°30′	4,650	0.5–3	4	0.1	Few nodules visible
Msn 8-29	N 20°38′	W 130°46′	5,180	1–6	20	1.2	
Naga 6–C	N 24°27′	W 135°17′	4,300	0.5–130	—	—	Nodules and pumice
Msn 8-24	N 13°04′	W 138°59′	4,990	0.5–15	—	—	Nodules and pumice
Msn 8-19	N 9°06′	W 145°18′	5,400	6–15	15	1.2	
Msn 8-11	N 2°00′	W 147°06′	4,480	—	1	—	Few scattered nodules
Msn H	N 10°02′	W 165°29′	1,650	1–50	10–60	1–5	Nodules and crusts
DWP 2	N 10°25′	W 130°35′	4,712	20	5–10	1	
DWP 3	N 3°12′	W 131°31′	4,440	—	—	—	No nodules visible

NODULE DISTRIBUTION AND CONCENTRATION

Station	Latitude	Longitude	Depth				Notes
DWP 4	N 1°21'	W 131°33'	4,510	—	—	—	No nodules visible
Msn 8-5	S 5°54'	W 149°38'	5,100	—	—	1	Thin sediment cover
Msn 8-1	S 13°53'	W 150°35'	3,695	2-5	65	1.5	Very uniform distribution
DWP 8	S 21°37'	W 147°40'	4,684	2-10	26	2.5	MENARD and SHIPEK (1958)
DWP 9	S 25°57'	W 146°21'	1,320	—	—	—	Nodules in ripples
DWP 10	S 32°08'	W 140°30'	4,770	—	23	1	MENARD and SHIPEK (1958)
DWP 11	S 42°50'	W 125°32'	4,560	1-10	46	1	Nodules and crusts
DWP 12	S 44°26'	W 110°40'	3,180	—	—	—	No nodules visible
Msn 6-25	S 54°35'	W 177°18'	5,330	—	—	—	Few nodules visible
Msn 6-2	S 61°40'	E 170°40'	5,160	2-8	80	2.5	Very uniform distribution
Msn P	S 5°05'	E 176°25'	5,342	—	—	—	Thin sediment cover
Msn S	S 9°00'	E 171°28'	5,000	2-5	50	1.5	
Msn U	S 12°35'	E 164°21'	4,250	—	—	0.5	Thin sediment cover
Msn W	S 14°47'	E 151°14'	4,400	1-4	—	1.0	Thin sediment cover

[1] For explanation of abbreviations of stations see Appendix I.

concentrations, the data were developed with which Fig. 62 was constructed. The curves of this figure show the maximum and minimum probable concentrations of nodules within an area in which a nodule of a given size has at least a 99% chance of being captured by a 5-cm diameter gravity core. Fig. 61 also locates those stations where gravity cores were known to have captured manganese nodules. Table XXIII lists the stations where nodule concentrations have been estimated from core-capture data. When a number of closely spaced cores were obtained in one locality, the concentration estimates of the stations in this group were averaged and the average of this group of stations was used to calculate the average of a region. This method was used to prevent any bias in the calculation of the overall regional average concentration by recording a large number of sample points within a limited area of unusually high or low concentrations of the nodules.

Russian oceanographers have used a method of determining sea-floor nodule concentrations by which a 0.25 m² area of the sea floor is removed with a clamshell-type bucket and taken to the surface for

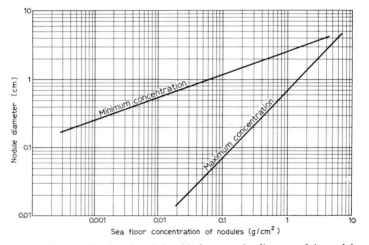

Fig. 62. Diagram, showing the relationship between the diameter of the nodules captured in a 5-cm-diameter sediment core and the maximum and minimum probable concentrations of the nodules on the ocean floor in the area where the nodules were captured.

TABLE XXIII

MANGANESE NODULE CONCENTRATIONS AT THE SURFACE OF THE PACIFIC OCEAN SEA-FLOOR SEDIMENTS AS DETERMINED BY CORING

Station	Location data		Depth (m)	Average nodule size (cm)	Surface concentration estimate[1] (g/cm²)	Sediment type
	Latitude	Longitude				
Eastern region						
Wig. 6	N 28°59'	W 125°41'	4,000	0.3	0.2	R. C.
DWHG 4	N 24°22'	W 125°00'	4,330	1 × 2 × 2	1.4	R. C.
Chub. 19	N 7°41'	W 125°37'	4,416	0.6	0.4	R. C.
Chub. 9	N 10°19'	W 125°27'	4,580	0.9	0.6	R. C.
Chub. 3	N 15°00'	W 125°26'	4,380	1	0.7	R. C.
Chub. 17	N 8°05'	W 125°25'	4,453	1	0.7	R. C.
Chub. 39	N 8°09'	W 125°02'	4,360	1	0.7	R. C.
Chub. 2	N 16°02'	W 125°01'	4,354	1 × 2 × 2	1.6	R. C.
Chub. 1	N 19°00'	W 121°53'	4,138	0.2	0.1	R. C.
Expl. 14b	N 19°46'	W 114°44'	3,438	1.5 × 3 × 2	1.8	R. C.
Expl. 14d	N 19°20'	W 114°12'	3,480	3 × 3 × 1	2.3	R. C.
Acap. 114	N 10°53'	W 105°07'	3,275	1	1.0	—
DWBG 147	N 1°27'	W 116°13'	4,000	1 × 1.5 × 1.5	1.0	Calc. O.
DWBG 19	S 14°59'	W 136°02'	4,465	0.5 × 1	0.5	R. C.
DWBG 18	S 13°37'	W 135°31'	4,337	1.5	1.1	R. C.
DWBG 17	S 12°51'	W 135°13'	4,318	0.5	0.4	R. C.
DWBG 16	S 6°05'	W 132°53'	4,855	0.5	0.5	Rad. O.

[1] Determined by taking an average of the maximum and minimum probable concentrations for the size of the nodule captured in the 5-cm diameter core as given in Fig. 15.

TABLE XXIII (continued)

Station	Location data		Depth (m)	Average nodule size (cm)	Surface concentration estimate (g/cm²)	Sediment type
	Latitude	Longitude				
Eastern region						
Cap. 33Bg	S 12°46'	W 143°33'	4,380	1	0.7	R. C.
PAS 19121	N 27°20'	W 116°10'	4,030	0.5 × 2 × 2	1.2	—
Msn 148G	N 9°06'	W 145°18'	5,400	1.4 × 2.5 × 2.7	1.5	R. C.
Msn 150G	N 10°59'	W 142°37'	4,978	1.1 × 0.8 × 1	0.4	R. C.
Msn 153PG	N 13°07'	W 138°56'	4,927	1 × 1 × 1	1.2	R. C.
Msn 157G	N 24°18'	W 126°30'	4,414	0.5 × 0.8 × 0.5	0.2	R. C.
DWBG 18	S 13°37'	W 135°31'	4,337	1	1.2	R. C.
Western region						
MP 43J	N 12°07'	E 164°52'	3,290	1.5	1.1	R. C.
Msn Q	S 7°03'	E 174°12'	5,378	1.5	1.2	R. C.
Cap. 30Bg	S 17°28'	W 160°59'	4,710	1	0.7	R. C.
Cap. 31Bg5	S 17°29'	W 158°40'	4,890	1	0.7	R. C.
Cap. H13	S 21°25'	E 177°46'	3,840	0.5	0.4	R. C.
Central region						
Cusp 8P	N 43°58'	W 140°38'	4,350	1.2	0.9	R. C.
Jyn II-21	N 36°29'	E 146°43'	5,720	2.5 × 1 × 1	0.7	R. C.
Cusp 15	N 37°15'	W 143°07'	5,220	2	1.7	R. C.
Tet. 22	N 16°06'	W 165°45'	2,400	2 × 2.5 × 3	2.1	R. C.
Tet. 24	N 15°02'	W 162°31'	5,666	2 × 1.3 × 1	0.7	R. C.
Tet. 27A	N 13°05'	W 163°10'	5,413	1.3 × 1 × 1	0.5	R. C.

NODULE DISTRIBUTION AND CONCENTRATION 171

Station	Latitude	Longitude	Depth	Size	Value	Type
Msn G	N 14°11'	W 161°08'	5,632	0.5 × 1 × 3	1.4	R. C.
Msn K	N 6°03'	W 169°59'	5,400	1.5 × 1.5	1.2	R. C.
Msn J	N 7°47'	W 168°00'	4,994	2 × 2.5 × 2.5	1.9	R. C.
Msn 128G	S 13°53'	W 150°35'	3,623	3 × 3 × 2	2.9	Coral S.
Msn 121G	S 29°35'	W 158°58'	5,252	1.5 × 1.5 × 2.5	1.2	R. C.
Msn 125G	S 26°01'	W 153°59'	5,038	3 × 3.5 × 4	3.0	R. C.
Msn 126G	S 24°41'	W 154°45'	4,542	1.1 × 3 × 3	2.6	R. C.
DWBG 37	S 29°09'	W 143°01'	4,120	2 × 4 × 4	3.8	R. C.
Msn 116P	S 35°50'	W 163°01'	4,950	2 × 2 × 1	2.0	R. C.
DWBG 44	S 34°25'	W 138°47'	4,860	0.5	0.8	R. C.
DWBG 47	S 36°33'	W 137°24'	4,700	2 × 1.5 × 2	0.9	R. C.
DWBG 46	S 36°23'	W 137°15'	4,680	1.5	1.1	R. C.
DWBG 48	S 37°05'	W 137°10'	4,940	1.5	1.1	R. C.
DWBG 52	S 40°36'	W 132°49'	5,120	3	3.0	Calc. O.
DWHG 31	S 35°11'	W 135°32'	4,700	1	0.7	—
DWBG 54	S 41°24'	W 129°06'	4,880	2–3	2.2	R. C.
DWHG 34	S 44°13'	W 127°20'	4,600	0.5–2.5	1.4	Calc. O.
DWBG 56	S 42°16'	W 125°50'	4,560	0.5–1.5	0.8	R. C.
DWBG 57A	S 42°50'	W 125°32'	4,560	0.5 × 2 × 2	1.5	Calc. O.
DWBG 58	S 43°07'	W 125°23'	4,640	2 × 3 × 3	2.6	R. C.
DWBG 59	S 44°23'	W 124°39'	4,500	2	1.7	R. C.
DWHG 48	S 42°00'	W 102°00'	4,240	3.5	3.8	R. C.
DWBG 78	S 44°08'	W 100°58'	4,100	1–2	1.2	Calc. O.
Msn 85G	S 57°43'	E 169°12'	5,288	2 × 2 × 3	1.5	R. C.
Msn 98P	S 54°31'	W 177°12'	5,274	1 × 1 × 1.7	0.6	R. C.
Msn 90G	S 63°04'	E 178°29'	3,583	2.5 × 3.5 × 1	2.5	Sil. O.
Msn 91G	S 64°11'	W 165°56'	2,932	2.3 × 2.5 × 2.5	2.5	Sil. O.

TABLE XXIV

SEA-FLOOR CONCENTRATIONS OF MANGANESE NODULES AS DETERMINED BY BOTTOM SCOOP METHODS

(After Skornyakova and Zenkevitch, 1961)

Station	Location data		Depth (m)	Percent of area covered by nodules (%)	Surface concentration estimate (g/cm^2)	Surface concentration estimate by photos at nearby station (g/cm^2)
	Latitude	Longitude				
Vit. 4243	N 24°56'	W 139°51'	4,368	4.1	0.05	—
Vit. 4245	N 25°00'	W 137°19'	4,645	20.1	0.18	—
Vit. 4273	N 19°59'	W 113°57'	3,820	6.2	0.11	0.60
Vit. 4285	N 19°57'	W 126°06'	4,576	25.0	0.23	0.36
Vit. 4289	N 20°00'	W 130°01'	5,005	6.7	0.11	—
Vit. 4343	N 24°00'	E 179°58'	5,815	22.2	0.40	—
Vit. 4347	N 24°00'	E 173°36'	5,318	50.0	0.60	—
Vit. 4351	N 23°57'	E 170°58'	5,817	10.0	0.17	—
Vit. 4355	N 24°02'	E 167°24'	6,052	16.0	0.60	—
Vit. 4359	N 24°01'	E 163°02'	5,542	36.0	1.00	1.90
Averages			5,096	20.0	0.35	

examination (SKORNYAKOVA and ZENKEVITCH, 1961). At the surface, the nodules from this chunk of the sea floor are removed and weighed. The details of the method are not known; however, if the sample device can capture all the nodules in the area of the sea floor covered by it and deliver them to the surface, highly accurate concentration measurements, over the rather small areas sampled, can be obtained. Table XXIV lists some stations where surface concentration measurements have been made using this method.

There appears to be no extensive area, other than the continental shelves and the deep ocean trenches, in the Pacific Ocean between latitudes 50° north and 60° south in which manganese nodules can not be found. AGASSIZ (1906) found nodules at almost every dredge station in the southeast Pacific Ocean along a tract that crossed this area in four widely-spaced traverses. MENARD and SHIPEK (1958) reported on a great abundance of nodules throughout the central part of the South Pacific. ZENKEVITCH and SKORNYAKOVA (1961) show a map of the general distribution of manganese nodules throughout the whole of the central Pacific Ocean.

If the data of Tables XXII, XXIII, and XXIV are plotted on a map of the Pacific Ocean, an increase in the concentration of the nodules as the central part of the Pacific is approached can be noticed. On the basis of the concentration data listed in Tables XXII, XXIII, and XXIV, the Pacific Ocean has been divided into three regions. These regions are shown in Fig. 61.

In the eastern Pacific, in an area of about 45 million km^2, 40 concentration estimates ranged between 0.05–2.3 g per cm^2 and averaged 0.78 g per cm^2. The 51 measurements in the central region ranged between 0.2–3.8 g per cm^2 and averaged 1.45 g per cm^2. Ten measurements in the western region ranged between 0.4–1.5 g per cm^2 and averaged 0.86 g per cm^2. The rate of sedimentation of the associated sediments and/or the activity of the agencies which keep the nodules at the sediment-water interface are thought to cause these regional-concentrational variations. The central region is probably a zone of generally low-sedimentation rates relative to the rate of growth of the nodules. Without the diluting effect of the associated sediments, the nodules can form more concentrated deposits.

Tonnage estimates

In Table XXV are listed statistics on the data of Tables XXII, XXIII, and XXIV. Also shown in Table XXV are the areas in the concentration regions shown in Fig. 61 with tonnage estimates of the individual regions. These tonnage estimates are based on the calculated average concentration of the nodules for a region prevailing over the whole of the region. While the estimates from photographs and cores yield estimates of approximately the same magnitude, the estimates for the grab samples were, on the average, about 30% of the average of the estimates by the other two methods. SKORNYAKOVA and ZENKEVITCH (1961) noticed this discrepancy and

TABLE XXV

STATISTICS ON THE DATA OF TABLES XXII, XXIII AND XXIV, AND TOTAL TONNAGES OF MANGANESE NODULES IN THE PACIFIC OCEAN

Statistic	Eastern region	Central region	Western region	Pacific Ocean
Number of photographs	11	13	5	29
Concentration estimates				
Maximum (g/cm^2)	1.2	2.5	1.5	2.5
Minimum (g/cm^2)	0.36	0.9	0.46	0.36
Average (g/cm^2)	0.86	1.60	0.90	0.97
Number of grab samples	5	5	0	10
Concentration estimates				
Maximum (g/cm^2)	0.23	1.00	—	1.00
Minimum (g/cm^2)	0.05	0.17	—	0.05
Average (g/cm^2)	0.14	0.56	—	0.35
Number of cores	24	33	5	62
Concentration estimates				
Maximum (g/cm^2)	2.3	3.8	1.2	3.8
Minimum (g/cm^2)	0.1	0.5	0.4	0.1
Average (g/cm^2)	0.89	1.71	0.82	1.32
Total of all measurements	40	51	10	101
Average concentrations of all methods (g/cm^2)	0.78	1.45	0.86	1.12
Area in region (km$^2 \times 10^6$)	44.9	62.1	47.2	154.2
Tonnage of nodules (billions of metric tons)	350	900	406	1,656

assumed that the small area of the grab sample and the small number of grab samples may have biased the estimates by this method.

As far as is known, there are no available data of a detailed exploration within a single deposit of the nodules, that is with sample points spaced a mile or so apart. Consequently, we cannot be certain of the continuity of the deposits of nodules around a given sample point in regard to concentration. Most of the calculations concerning manganese-nodule tonnages are of the nature of speculation. The uniform character of pelagic sedimentation over extensive areas would lead one to suspect that the nodule deposits are widespread and that they are relatively uniform both in concentration and composition.

By using the data of Table XXII, XXIII, and XXIV and taking areas from Fig. 61, the tonnage of manganese nodules at the surface of the sediments of the Pacific Ocean is indicated to be $1.66 \cdot 10^{12}$ metric tons. The tonnage figures for the individual regions of the Pacific Ocean are given in Table XXV. The tonnage calculation for the central region is probably conservative as manganese crusts were not taken into account. If sufficient data allowed the crusts to be included in quantity calculations, tonnage estimates would greatly increase. The concentration of manganese dioxide formed by a 5-cm thick crust, for example, would be 12.0 g per cm^2 assuming a specific gravity for the crust of 2.4.

MENARD AND SHIPEK (1958) estimated a total of 10^{11} tons of nodules in 10^7 km^2 of the southwestern Pacific basin. This figure compares favorably with my estimate of $17 \cdot 10^{11}$ tons for $17 \cdot 10^7$ km^2 of the whole of the Pacific Ocean. ZENKEVITCH and SKORNYAKOVA (1961), however, estimate a total Pacific Ocean surficial tonnage of $0.9 \cdot 10^{11}$, which is about one-twentieth of my estimate.

Rate of accumulation of manganese nodules

The area of the ocean floor physically occupied by the nodules, in areas where they have been photographed, averages about 20% (MENARD and SHIPEK, 1958; SKORNYAKOVA and ZENKEVITCH, 1961). Assuming that the rate of formation of the nodules is, on the average, about 0.1 mm per 1,000 years, the nodules are indicated to be forming at the rate of $6 \cdot 10^6$ metric tons per year in the Pacific Ocean. At

TABLE XXVI

CHEMICAL ANALYSES OF MANGANESE NODULES RECOVERED BY THE "CHALLENGER" EXPEDITION

"Challenger" station	Longitude	Latitude	Depth (m)	Weight percent (Total weight of sample basis)[1]									Percent soluble in HCl
				MnO_2	Fe_2O_3	SiO_2	Al_2O_3	$CaCO_3$	$CaSO_4$	$Ca_3(PO_4)_2$	$MgCO_3$	H_2O	
3	W 20°14'	N 25°45'	3,620	25.6	32.9	6.0	3.5	3.5	1.2	0.9	1.6	24.8	70.5
3a	W 20°14'	N 25°45'	3,620	22.8	41.4	6.7	2.3	5.4	1.2	0.3	1.7	18.3	78.4
16	W 50°33'	N 20°39'	4,450	29.3	26.6	8.8	4.0	2.2	1.1	traces	4.5	13.6	82.7
160a[2]	E 134°10'	S 42°42'	4,750	32.5	19.6	19.0	2.5	3.4	0.6	0.2	1.9	20.4	70.3
160b	E 134°10'	S 42°42'	4,750	39.3	17.5	21.8	4.8	3.3	0.6	traces	1.7	11.0	75.6
160c	E 134°10'	S 42°42'	4,750	33.6	16.6	28.3	3.9	3.4	0.6	traces	3.3	10.3	73.7
248	E 161°52'	N 37°41'	5,300	22.5	21.7	29.1	4.7	3.3	0.9	traces	1.4	16.5	61.1
252a	W 160°17'	N 37°52'	5,020	28.5	20.5	27.7	5.9	3.8	0.9	traces	2.1	10.6	68.5
252b	W 160°17'	N 37°52'	5,020	25.4	18.7	21.3	6.7	4.1	0.6	traces	2.5	20.8	64.5
252c	W 160°17'	N 37°52'	5,020	25.5	17.8	27.6	6.6	4.2	0.6	traces	2.5	15.2	62.5
253	W 156°25'	N 38°09'	5,720	26.2	22.1	27.1	6.5	3.6	0.8	0.5	1.2	12.1	73.2
256	W 154°56'	N 30°22'	5,400	39.6	19.6	17.8	3.7	2.9	0.6	traces	4.5	11.3	77.6
264	W 152°37'	N 14°19'	5,480	29.1	23.1	28.3	3.3	3.0	0.6	traces	3.7	8.9	83.9
274a	W 152°15'	S 7°25'	5,030	51.5	9.0	13.5	1.6	4.0	0.6	1.4	5.1	12.6	82.6
274b	W 152°15'	S 7°25'	5,030	52.4	12.9	12.5	0.9	4.3	0.8	0.9	2.2	12.5	82.9
274c	W 152°15'	S 7°25'	5,030	55.9	10.5	11.3	0.6	4.2	0.6	0.4	4.3	11.4	84.5
276	W 149°30'	S 13°28'	4,300	11.4	42.0	15.5	6.3	5.8	0.9	traces	1.8	16.3	77.6
281a	W 150°11'	S 22°21'	4,360	22.2	30.3	22.6	3.3	3.6	0.3	traces	1.7	16.0	73.2
281b	W 150°11'	S 22°21'	4,360	19.9	34.0	24.4	4.7	3.7	0.6	—	1.8	11.0	75.8
285a	W 137°43'	S 32°36'	4,350	36.5	25.4	16.8	4.4	2.4	0.3	traces	1.2	12.9	76.2
285b	W 137°43'	S 32°36'	4,350	24.7	13.8	27.7	11.2	4.0	0.7	7.2	1.4	9.3	69.2

NODULE DISTRIBUTION AND CONCENTRATION 177

285c	W 137°43'	S 32°36'	4,350	16.1	23.1	24.7	8.6	6.3	0.9	traces	1.1	19.3	60.5
285d	W 137°43'	S 32°36'	4,350	22.1	17.5	26.0	14.2	2.4	1.1	2.6	1.2	13.0	63.1
285e	W 137°43'	S 32°36'	4,350	22.2	14.2	21.8	11.5	5.1	0.8	0.9	0.1	23.4	58.5
286a	W 133°22'	S 33°29'	4,270	27.4	25.2	26.0	4.4	5.2	0.9	0.7	1.5	8.7	78.2
286b	W 133°22'	S 33°29'	4,270	22.8	24.1	28.4	4.9	3.2	0.5	0.7	1.0	15.5	65.6
286c	W 133°22'	S 33°29'	4,270	38.2	17.9	20.0	2.8	5.4	0.9	traces	3.5	11.4	75.0
289	W 131°23'	S 39°41'	4,660	32.0	21.0	21.5	4.2	3.9	0.6	0.4	2.1	13.8	69.7
293	W 105°05'	S 39°04'	3,740	37.6	20.9	16.7	3.2	5.0	0.7	0.7	4.0	11.2	76.2
297	W 83°07'	S 37°29'	3,250	30.8	29.8	14.7	1.0	7.0	0.9	traces	4.6	11.3	81.6
299a	W 74°43'	S 33°31'	3,950	55.7	6.8	14.1	3.0	6.1	0.6	traces	2.0	11.8	77.5
299b	W 74°43'	S 33°31'	3,950	46.9	14.7	17.4	2.9	3.1	0.6	traces	4.4	10.0	77.5
299c	W 74°43'	S 33°31'	3,950	63.2	6.5	11.1	2.4	3.1	0.5	traces	2.8	10.4	80.6
302	W 82°11'	S 42°43'	2,650	22.3	40.9	15.0	1.2	4.5	1.3	traces	3.6	11.4	82.8

[1] Wet chemical assays (After MURRAY and RENARD, 1891). $CaCO_3$ = sum of $CaCO_3$ and CaO. $MgCO_3$ = sum of $MgCO_3$ and MgO. H_2O determined by drying for one hour at 110°C.
[2] Analyses on different nodules from the same dredge-haul station.

this rate, about 200,000 years would be required for the formation of the nodules now speculated to be present at the surface of the Pacific Ocean pelagic sediments.

The red clays are, in general, the slowest forming of the pelagic sediments and are also the sediments most commonly associated with the manganese nodules. At an average rate of formation of the red clays of 0.1 cm per 1,000 years (GOLDBERG and ARRHENIUS, 1958), about 20 cm of this sediment should have formed over the past 200,000 years. If this rate of formation of the red clays prevailed uniformly over the whole of the Pacific Ocean, and no agency was at work to keep the nodules at the surface of the sediments, the nodules should be buried. If the nodules maintained a growth rate equal to that of their associated sediments, they should be, on the average, much greater in size than they have been found to be. In fact, they should be, on the average, at least 40 cm in diameter, whereas their average diameter does not, in general, exceed about 4 cm. At an average growth rate of 0.1 mm per 1,000 years, the time necessary for a nodule to accrete to a diameter of 4 cm should be about 800,000 years assuming accretion is taking place, on the average, over one half the surface area of the nodule at any given time. This discrepancy between the average size of the nodules and the rates of formation of the associated sediments is further magnified when considering the nodules found associated with the calcareous oozes as, in general, the oozes form more rapidly than the clays and when considering that the assumed growth rate of 0.1 mm per 1,000 years for the nodules may be optimistic by a factor of about 10.

Unless, therefore, the nodule deposits are much less extensive than is suspected or unless the nodules are forming at a higher rate than has been measured, some agency must be at work to maintain the nodules at the surface of their associated sediments.

CHEMICAL COMPOSITION OF MANGANESE NODULES

Table XXVI lists analyses on 34 samples of manganese nodules from eighteen different locations in the oceans. Two of these samples are from the Atlantic Ocean, one from the Indian Ocean, and fifteen from the Pacific Ocean. The assays listed in Table XXVI are for major constituents only and the compositional formulae used in this table do not necessarily describe the chemical compounds actually found in the nodules. A number of assays listed in Table XXVI were obtained by analyzing different nodules from the same dredge haul or by analyzing different parts, such as the shell or the nucleus, of the same nodule, and, thus, yield some information concerning compositional variations within a single deposit of the nodules. Table XXVII lists statistics on the data of Table XXVI, including average assays for the major constituents of the manganese nodules.

Table XXVIII lists the maximum, minimum, and average weight percentages of 27 of the elements found in the ocean-floor manganese nodules. The assays which yielded the data of Table XXVIII were taken from 54 widely separated (500–1,000 km) locations so as to yield as true a statistical picture on the ocean-wide composition of the nodules from these oceans as possible. The elements of major compositional importance in the nodules not listed in Table XXVIII are oxygen and carbon. Oxygen generally constitutes about 50% of the bulk weight of the manganese nodules. Concentrations of

TABLE XXVII

MAXIMUM, MINIMUM, AND AVERAGE PERCENTAGES OF THE MAJOR CONSTITUENTS OF OCEAN-FLOOR MANGANESE NODULES[1]

Material	Weight percentages[2]		
	maximum	minimum	average
MnO_2	63.2	11.4	31.7
Fe_2O_3	42.0	6.5	24.3
SiO_2	29.1	6.0	19.2
Al_2O_3	14.2	0.6	3.8
$CaCO_3$	7.0	2.2	4.1
$CaSO_4$	1.3	0.3	0.8
$Ca_3(PO_4)_2$	1.4	traces	0.3
$MgCO_3$	5.1	0.1	2.7
H_2O	24.8	8.7	13.0
Insoluble in HCl	38.9	16.1	26.8

[1] Statistics derived from data of Table XXVI.
[2] On a total weight of air-dried sample basis.

certain minor constituents in the manganese nodules are listed in Table XXIX.

Any comprehensive theory to explain the formation of the man-

TABLE XXVIII

MAXIMUM, MINIMUM, AND AVERAGE WEIGHT PERCENTAGES OF 27 ELEMENTS IN MANGANESE NODULES FROM THE PACIFIC AND ATLANTIC OCEANS

Element	Weight percentages (dry-weight basis)[1]					
	Pacific Ocean, statistics on 54 samples			Atlantic Ocean, statistics on 4 samples		
	maximum	minimum	average	maximum	minimum	average
B	0.06	0.007	0.029	0.05	0.009	0.03
Na	4.7	1.5	2.6	3.5	1.4	2.3
Mg	2.4	1.0	1.7	2.4	1.4	1.7
Al	6.9	0.8	2.9	5.8	1.4	3.1
Si	20.1	1.3	9.4	19.6	2.8	11.0
K	3.1	0.3	0.8	0.8	0.6	0.7
Ca	4.4	0.8	1.9	3.4	1.5	2.7
Sc	0.003	0.001	0.001	0.003	0.002	0.002
Ti	1.7	0.11	0.67	1.3	0.3	0.8
V	0.11	0.021	0.054	0.11	0.02	0.07
Cr	0.007	0.001	0.001	0.003	0.001	0.002
Mn	41.1	8.2	24.2	21.5	12.0	16.3
Fe	26.6	2.4	14.0	25.9	9.1	17.5
Co	2.3	0.014	0.35	0.68	0.06	0.31
Ni	2.0	0.16	0.99	0.54	0.31	0.42
Cu	1.6	0.028	0.53	0.41	0.05	0.20
Zn	0.08	0.04	0.047	—	—	—
Ga	0.003	0.0002	0.001	—	—	—
Sr	0.16	0.024	0.081	0.14	0.04	0.09
Y	0.045	0.016	0.033	0.024	0.008	0.018
Zr	0.12	0.009	0.063	0.064	0.044	0.054
Mo	0.15	0.01	0.052	0.056	0.013	0.035
Ag	0.0006	—	0.0003[2]	—	—	—
Ba	0.64	0.08	0.18	0.36	0.10	0.17
La	0.024	0.009	0.016	—	—	—
Yb	0.0066	0.0013	0.0031	0.007	0.002	0.004
Pb	0.36	0.02	0.09	0.14	0.08	0.10
L.O.I.[3]	39.0	15.5	25.8	30.0	17.5	23.8

[1] As determined by X-ray emission spectrography.
[2] Average of 5 samples in which Ag was detected.
[3] L.O.I. = Loss on ignition at 1100°F for one hour. The L.O.I. figures are based on a total weight of air-dried sample basis.

TABLE XXIX

CONCENTRATIONS OF CERTAIN MINOR CONSTITUENTS IN THE MANGANESE NODULES

Element	Number of analyses	Weight percentages			Concentration of element in sea water[1] (p.p.b.)	Concentration ratio[2]	Reference
		maximum (p.p.m.)	minimum (p.p.m.)	average (p.p.m.)			
Be	7	5	2	3	0.03	10^5	Riley and Sinhaseni (1958)
Ge	2	6	5	6	0.06	10^5	
Nb	8	150	30	85	—	—	Riley and Sinhaseni (1958)
Cd	2	11	6	9	0.1	10^5	Riley and Sinhaseni (1958)
Sn	3	320	240	267	3	10^5	Riley and Sinhaseni (1958)
Ce	8	300	30	134	0.4[3]	$3 \cdot 10^5$	
Ni	7	300	150	280	—	—	
W	2	95	47	66	0.1	$7 \cdot 10^5$	Riley and Sinhaseni (1958)
Hg	1	2	2	2	0.03	10^5	Riley and Sinhaseni (1958)
Tl[4]	7	300	150	170	0.01	$2 \cdot 10^7$	
Tl[4]	3	110	80	93	0.01	—	Riley and Sinhaseni (1958)
Bi	12	45	22	30	0.2	10^5	
Ra	—	$80 \cdot 10^{-6}$	$1 \cdot 10^{-6}$	—	—	—	Riley and Sinhaseni (1958)
Th	—	143	24	50	0.05	10^6	Riley and Sinhaseni (1958)
U	5	5.0	3.6	4.2	3	10^3	Tatsumoto and Goldberg (1959)

[1] Data for concentrations of elements in sea water from Riley and Sinhaseni (1958). p.p.b. = parts per billion.
[2] Ratio of the concentration of the element in the nodules to that of the element in sea water.
[3] Value from Sverdrup et al., (1942, p. 220).
[4] Because the data from the present author's work differ quite considerably from those obtained by Riley and Sinhaseni (1958), both are included in this table.

TABLE XXX

Station	Vit. 3150	Jyn II–8	Vit. 4074	Vit. 4104
Latitude	N 44°28'	N 40°29'	N 40°24'	N 41°08'
Longitude	E 170°15'	E 172°33'	W 175°42'	W 159°54'
Depth (m)	1,258	4,250	6,065	5,445
Asso. sed.[2]	Volc. R.[2]	R. C.	R. C.	Volc. R.
Sampler	Trawl	Core	Trawl	Trawl
Nod. dia. (cm)	?–13	1–cm	1–12	1–4
Sp. G.	—	—	—	—
Sect. sampled	Out. 1 cm	Whole nod.	Out. layer	Out. layer
Chemical analyses (Weight percentages)				
Al_2O_3	1.04	6.6	8.21	7.63
SiO_2	6.09[6]	39.6	30.87	31.59
P	—[3]	—	—	0.14
K	1.27	—	1.01	—
Ca	1.90	0.91	1.87	1.93
Ti	0.70	0.28	0.40	0.44
Mn	33.9	7.8	12.0	13.2
Fe	7.8	12.4	10.7	8.9
Co	0.41	0.48	0.13	0.31
Ni	0.42	0.20	0.22	0.43
Cu	—	0.21	—	—
Zn	—	0.056	—	—
Sr	—	0.11	—	—
Mo	—	0.000	—	—
Ba	0.92	0.30	—	—
Pb	—	0.13	—	—
H_2O[4]	15.50	11.0	15.92	
L.O.I.[5]	—	—	—	20.14
Reference	Skornyakova et al. (1962)		Skornyakova et al. (1962)	Skornyakova et al. (1962)
Reduced analyses				
Mn	43.7	18.2	26.7	32.4
Fe	10.1	29.0	23.8	21.9
Co	0.53	1.12	0.29	0.76
Ni	0.54	0.47	0.49	1.1
Cu	—	0.49	—	—
Pb	—	0.13	—	—

[1] Chemical analyses on an air-dried as received by X-ray fluorescence spectrography. Analyses referenced to other investigators were performed by the following techniques: Willis and Ahrens (1962) and Goldberg and Menard (pers. comm., 1960), emission spectrography; Murray and Renard (1891), wet chemical; Riley and Sinhaseni (1958), various methods, but mainly by spectrographic techniques; Skornyakova et al. (1962) wet chemical and colorimetric. The phosphorus assays were by wet chemical methods.

[2] Asso. sed. — R. C. = red clay; Calc. O. = calcareous oozes; T. Sed. = terrigenous sediments; Volc. R. = volcanic rock; Coral D. = coral debris; Rad. O. = radiolarian ooze; Coral S. = coral sand; G. O. = *Globigerina* ooze;

CHEMICAL COMPOSITION OF PACIFIC OCEAN MANGANESE NODULES[1]

NH C-10	Cusp 8P	Vit. 4191	Vit. 4191c	Station
N 40°14'	N 43°58'	N 40°20'	N 40°20'	Latitude
W 155°06'	W 140°38'	W 135°47'	W 135°47'	Longitude
5,500	4,350	4,560	4,560	Depth (m)
R. C.	R. C.	R. C.	R. C.	Asso. sed.[2]
Core wire	Core	—	Trawl	Sampler
61×61×32	1.2	3	3–10	Nod. dia. (cm)
2.19	—	2.41	—	Sp. G.
Out. 2 cm	x-sect.	x-sect.	Out. layer	Sect. sampled
			Chemical analyses (Weight percentages)	
12.0	7.1	8.1	4.97	Al_2O_3
29.9	25.2	28.5	22.05	SiO_2
—	—	—	0.18	P
1.71	0.54	0.73	—	K
1.00	1.73	1.43	1.86	Ca
0.63	0.55	0.49	0.62	Ti
11.9	17.7	16.5	15.2	Mn
6.9	9.4	9.5	12.5	Fe
0.23	0.23	0.22	0.35	Co
0.45	0.72	0.58	0.37	Ni
0.47	0.42	0.36	—	Cu
0.057	0.091	0.076	—	Zn
0.077	0.091	0.070	—	Sr
0.038	0.036	0.026	—	Mo
0.57	0.43	0.36	—	Ba
0.20	0.13	0.13	—	Pb
19.3	18.5	19.2	—	H_2O[4]
—	—	—	20.14	L.O.I.[5]
			Skornyakova et al. (1962)	Reference
				Reduced analyses
30.7	36.0	37.3	31.0	Mn
17.8	19.1	21.5	25.6	Fe
0.59	0.47	0.50	0.72	Co
1.16	1.46	1.31	0.76	Ni
1.21	0.85	0.82	—	Cu
0.52	0.26	0.29	—	Pb

Gr. Mud. = green mud; Sil. O. = siliceous ooze.

[3] A dash, (—), indicated no determination was made.

[4] H_2O — determined by drying air-dried samples of the nodules at 200° Centigrade for two hours. In the Murray and Renard (1891) analyses, the H_2O was determined by drying samples of the nodules for several hours at 110°C.

[5] L.O.I. = loss on ignition, determined by drying air-dried samples of the nodules at 1100°F for one hour.

[6] In reducing the analyses to a detrital mineral free basis, SiO_2, Al_2O_3, H_2O, and $CaCO_3$ in amounts greater than 5%, were considered as detrital minerals.

TABLE

Station	Fan BD-20	Fan BD-25	Japan A	Japan B
Latitude	N 40°15′	N 40°23′	N 33°51′	N 34°23′
Longitude	W 128°27′	W 127°59′	E 138°41′	E 139°05′
Depth (m)	5,400	1,260	110	260
Asso. sed.	—	Basalt R.	T. Sed.	T. Sed.
Sampler	Dredge	Dredge	Dredge	Dredge
Nod. dia. (cm)	3–20	1 cm Crust	—	—
Sp. G.	—	—	2.89	—
Sect. sampled	x-sect.	x-sect.	x-sect.	x-sect.

Chemical analyses (Weight percentages)

Al_2O_3	6.8	4.7	0.6	2.27
SiO_2	27.2	15.2	0.5	0.86
P	—	—	—	—
K	—	—	0.6	0.5
Ca	1.09	1.21	28.8	4.0
Ti	0.18	0.21	0.002	0.05
Mn	18.8	23.5	12.6	37.0
Fe	8.4	8.8	0.77	0.6
Co	0.07	0.43	0.006	0.092
Ni	0.56	0.61	0.042	0.051
Cu	0.37	0.04	0.007	0.010
Zn	0.13	0.09	0.007	0.03
Sr	0.08	0.08	0.35	0.16
Mo	0.03	0.028	0.007	0.022
Ba	0.23	0.48	0.59	1.5
Pb	0.09	0.11	0.015	0.05
H_2O	11.8	13.3	6.6	—
L.O.I.	—	—	—	16
Reference				GOLDBERG and MENARD (pers. comm., 1960)

Reduced analyses

Mn	42.5	35.2	49.8	46.3
Fe	19.0	13.2	3.0	0.75
Co	0.16	0.65	0.023	0.11
Ni	1.27	0.91	0.16	0.063
Cu	0.86	0.06	0.028	0.012
Pb	0.29	0.13	0.059	0.062

XXX (continued)

Jyn II-21	JEDS 5	Chal. 248	Vit. 4084	Station
N 36°29'	N 38°00'	N 37°41'	N 35°00'	Latitude
E 146°43'	E 146°00'	W 177°04'	W 172°57'	Longitude
5,720	3,500	5,310	5,971	Depth (m)
R. C.	—	R. C.	R. C.	Asso. sed.
Core	Trawl	Trawl	Spoon	Sampler
2.5×1×1	3 cm Crust	7	3–4	Nod. dia. (cm)
—	—	—	—	Sp. G.
Whole nod.	x-sect.	Whole nod.	Out. 0.5 cm	Sect. sampled

				Chemical analyses (Weight percentages)
11.9	1.17	7.4	6.85	Al_2O_3
56.5	7.0	24.1	27.72	SiO_2
—	—	0.14	0.15	P
—	—	1.22	—	K
1.31	2.12	1.24	1.55	Ca
0.18	0.50	0.34	0.41	Ti
1.9	19.8	16.5	13.1	Mn
11.8	13.9	10.3	10.8	Fe
0.001	0.29	0.090	0.22	Co
0.12	0.38	0.28	0.29	Ni
0.07	0.10	0.43	—	Cu
0.041	0.057	0.38	—	Zn
0.057	0.13	0.10	—	Sr
0.000	0.048	0.027	—	Mo
0.05	0.72	0.36	—	Ba
0.08	0.15	0.094	—	Pb
5.3	21.7	17.0	—	H_2O
—	—	—	20.77	L.O.I.
		Riley and Sinhaseni (1958)	Skornyakova et al. (1962)	Reference

				Reduced analyses
7.2	28.2	32.0	29.3	Mn
45.0	19.7	20.0	24.2	Fe
0.004	0.41	0.18	0.49	Co
0.46	0.54	0.54	0.65	Ni
0.27	0.11	0.85	—	Cu
0.15	0.08	0.18	—	Pb

TABLE

Station	Vit. 4090c	Chal. 252	Chal. 253	Chal. 256
Latitude	N 35°02'	N 37°52'	N 38°09'	N 30°22'
Longitude	W 166°28'	W 160°17'	W 156°25'	W 154°56'
Depth (m)	5,907	5,020	5,720	5,400
Asso. sed.	R. C.	R. C.	R. C.	R. C.
Sampler	Spoon	Trawl	Dredge	Dredge
Nod. dia. (cm)	3–12	$6 \times 6 \times 8$	$6 \times 20 \times 31$	$3 \times 4 \times 3$
Sp. G.	—	—	—	—
Sect. sampled	Out. 0.8 cm	Whole nod.	—	x-sect.
Chemical analyses (Weight percentages)				
Al_2O_3	5.97	4.26	6.4	6.7
SiO_2	22.88	18.8	23.4	31.2
P	0.18	0.116	0.09	—
K	—	1.28	—	—
Ca	1.64	0.99	2.0	1.17
Ti	0.50	0.59	—	0.39
Mn	15.9	19.9	16.6	13.9
Fe	11.2	12.4	15.5	10.4
Co	0.36	0.13	—	0.31
Ni	0.61	0.40	—	0.68
Cu	—	0.26	—	0.43
Zn	—	0.31	—	0.06
Sr	—	0.057	—	0.08
Mo	—	0.037	—	0.018
Ba	—	0.34	—	0.45
Pb	—	0.15	—	0.16
H_2O	—	17.9	12.1	10.2
L.O.I.	22.80	—	—	—
Reference	SKORNYA-KOVA et al. (1962)	RILEY and SINHASENI (1958)	MURRAY and RENARD (1891)	
Reduced analyses				
Mn	32.8	33.7	27.2	26.8
Fe	23.2	21.1	25.3	20.0
Co	0.74	0.22	—	0.60
Ni	1.26	0.68	—	1.31
Cu	—	0.44	—	0.83
Pb	—	0.25	—	0.12

XXX (continued)

UPWD 2	UPWD 1	Vit. 4199	SOB 22	Station
N 34°08'	N 34°04'	N 35°07'	N 31°21'	Latitude
W 145°57'	W 145°56'	W 137°53'	W 119°03'	Longitude
5,300	5,390	5,035	915	Depth (m)
R. C.	R. C.	R. C.	T. Sed.	Asso. sed.
Dredge	Dredge	—	Dredge	Sampler
1–5	2–7	3–9	Crust	Nod. dia. (cm)
—	—	2.47	2.49	Sp. G.
x-sect.	x-sect.	x-sect.	x-sect.	Sect. sampled
				Chemical analyses (Weight percentages)
7.0	6.7	9.3	11.3	Al_2O_3
24.2	24.2	28.6	26.6	SiO_2
—	—	—	—	P
—	—	1.16	1.14	K
1.24	1.14	1.14	0.71	Ca
0.48	0.39	0.65	0.34	Ti
15.0	15.7	10.4	11.7	Mn
11.6	11.8	13.0	10.3	Fe
0.34	0.34	0.29	0.19	Co
0.59	0.67	0.33	0.24	Ni
0.34	0.45	0.29	0.06	Cu
0.048	0.060	0.062	0.048	Zn
0.093	0.080	0.076	0.055	Sr
0.027	0.03	0.017	0.036	Mo
0.58	0.50	0.49	0.47	Ba
0.16	0.17	0.19	0.091	Pb
14.8	16.0	14.6	9.9	H_2O
—	—	—	—	L.O.I.
				Reference
				Reduced analyses
27.8	29.6	21.9	22.4	Mn
21.5	22.2	27.4	19.7	Fe
0.63	0.64	0.61	0.36	Co
1.10	1.26	0.70	0.46	Ni
0.63	0.85	0.61	0.12	Cu
0.09	0.11	0.40	0.17	Pb

TABLE

Station	S Clem SV	SOB-20	SOB 5	SOB 10
Latitude	N 32°45′	N 31°23′	N 31°19′	N 30°12′
Longitude	W 118°13′	W 118°03′	W 117°38′	W 117°38′
Depth (m)	1,588	1,040	2,110	1,300
Asso. sed.	T. Sed.	T. Sed.	T. Sed.	T. Sed.
Sampler	Dredge	Dredge	Dredge	Dredge
Nod. dia. (cm)	15	Crust	Crust	Crugt
Sp. G.	2.26	2.35	2.34	—
Sect. sampled	x-sect.	x-sect.	x-sect.	x-sect.
Chemical analyses (Weight percentages)				
Al_2O_3	4.3	5.1	7.4	6.6
SiO_2	21.0	21.0	26.8	23.3
P	—	—	—	—
K	0.34	0.42	0.92	0.55
Ca	1.62	1.65	0.94	1.48
Ti	0.40	0.63	0.25	0.53
Mn	14.5	13.7	13.4	10.7
Fe	16.1	14.5	11.4	14.7
Co	0.14	0.53	0.083	0.40
Ni	0.19	0.23	0.34	0.18
Cu	0.052	0.052	0.061	0.035
Zn	0.048	0.040	0.067	0.036
Sr	0.15	0.15	0.074	0.13
Mo	0.060	0.048	0.036	0.041
Ba	0.48	0.61	0.41	0.46
Pb	0.14	0.21	0.075	0.24
H_2O	23.7	20.6	15.4	19.0
L.O.I.	—	—	—	—
Reference				
Reduced analyses				
Mn	28.4	25.7	26.6	21.0
Fe	31.6	27.2	22.6	28.8
Co	0.27	1.00	0.17	0.78
Ni	0.37	0.43	0.67	0.35
Cu	0.10	0.098	0.12	0.069
Pb	0.27	0.39	0.15	0.47

XXX (continued)

Vit. 4370	Vit. 4370	Vit. 4362	Vit. 4359	*Station*
N 26°12′	N 26°12′	N 24°04′	N 24°01′	*Latitude*
E 153°44′	E 153°44′	E 160°46′	E 163°02′	*Longitude*
6,120	6,120	3,951	5,542	*Depth (m)*
R. C.	R. C.	Calc. O.	R. C.	*Asso. sed.*
Trawl	Trawl	Spoon	Spoon	*Sampler*
1–5	1–5	1–6	1–9	*Nod. dia. (cm)*
—	—	—	—	*Sp. G.*
Out. layer	Whole nod.	—	Core of nod.	*Sect. sampled*
			Chemical analyses (Weight percentages)	
5.88	7.2	3.65	7.42	Al_2O_3
15.34	17.5	13.10	20.30	SiO_2
0.22	—	—	—	P
—	0.5	0.84	0.96	K
1.57	1.5	2.80	2.36	Ca
0.70	0.6	0.67	0.73	Ti
16.1	12.2	22.5	17.2	Mn
14.6	14.0	11.3	11.2	Fe
0.36	0.14	0.47	0.32	Co
0.41	0.41	0.49	0.54	Ni
—	0.27	—	—	Cu
—	—	—	—	Zn
—	0.06	—	—	Sr
—	0.03	—	—	Mo
	0.10			Ba
—	0.06			Pb
—	—	20.59	18.57	H_2O
26.99	23.0	—	—	L.O.I.
Skornya-	Goldberg	Skornya-	Skornya-	*Reference*
kova et al.	and Menard	kova et al.	kova et al.	
(1962)	(pers. comm.,	(1962)	(1962)	
	1960)			
				Reduced analyses
31.1	21.0	42.7	31.4	Mn
28.2	24.0	21.4	20.5	Fe
0.70	0.24	0.89	0.58	Co
0.79	0.70	0.93	0.99	Ni
—	0.46	—	—	Cu
—	0.10	—	—	Pb

TABLE

Station	Vit. 4351	Vit. 4331	Naga 16	Naga 15
Latitude	N 23°57′	N 20°03′	N 22°00′	N 23°54′
Longitude	E 170°58′	W 171°38′	W 150°00′	W 148°00′
Depth (m)	5,817	3,477	5,240	5,220
Asso. sed.	R. C.	Calc. O	R. C.	R. C.
Sampler	Spoon	Spoon	Core	Dredge
Nod. dia. (cm)	1–5	1–3	$2 \times 1.5 \times 1.5$	$1 \times 3 \times 3$
Sp. G.	—	—	2.50	2.59
Sect. sampled	Out. layer	Out. layer	x-sect.	x-sect.
Chemical analyses (Weight percentages)				
Al_2O_3	4.98	2.84	6.5	14.9
SiO_2	12.97	14.20	13.9	29.6
P	0.17	—	—	—
K	—	0.55	0.54	2.24
Ca	1.76	2.33	1.50	0.63
Ti	0.60	1.11	2.52	0.68
Mn	20.2	26.8	15.5	10.9
Fe	11.9	17.7	14.2	7.2
Co	0.46	0.46	0.37	0.20
Ni	0.53	0.27	0.46	0.49
Cu	—	—	0.33	0.43
Zn	—	—	0.067	0.060
Sr	—	—	0.11	0.060
Mo	—	—	0.034	0.035
Ba	—	—	0.71	0.35
Pb	—	—	0.13	0.12
H_2O	—	21.57	21.3	16.7
L.O.I.	27.00	—	—	—
Reference	Skornya-kova et al. (1962)	Skornya-kova et al. (1962)		
Reduced analyses				
Mn	36.7	43.6	26.6	28.1
Fe	21.6	28.8	24.4	18.6
Co	0.83	0.75	0.64	0.53
Ni	0.96	0.44	0.79	1.26
Cu	—	—	0.57	1.11
Pb	—	—	0.22	0.31

XXX (continued)

Vit. 4239	Naga 10C	Naga 8C	Vit. 4289	Station
N 24°50'	N 23°17'	N 23°17'	N 20°00'	Latitude
W 144°05'	W 141°13'	W 138°15'	W 130°01'	Longitude
5,190	5,540	4,890	4,895	Depth (m)
R. C.	R. C.	R. C.	R. C.	Asso. sed.
—	Core	Dredge	Dredge	Sampler
—	0.5×3	—	2×1×1	Nod. dia. (cm)
2.32	2.54	—	2.51	Sp. G.
x-sect.	x-sect.	x-sect.	x-sect.	Sect. sampled
			Chemical analyses (Weight percentages)	
17.4	5.6	4.7	7.3	Al_2O_3
38.7	12.6	14.0	15.6	SiO_2
—	—	—	—	P
2.60	0.53	0.62	0.92	K
0.69	1.66	1.35	1.51	Ca
0.45	1.40	1.61	0.65	Ti
7.0	19.1	15.9	21.8	Mn
4.7	12.3	14.1	8.7	Fe
0.14	0.46	0.52	0.33	Co
0.44	0.45	0.35	1.10	Ni
0.45	0.32	0.20	0.91	Cu
0.038	0.043	0.064	0.093	Zn
0.055	0.12	0.11	0.089	Sr
0.017	0.046	0.028	0.046	Mo
0.16	0.88	0.79	0.60	Ba
0.16	0.16	0.17	0.11	Pb
14.0	20.2	21.9	12.9	H_2O
—	—	—	—	L.O.I.
				Reference
				Reduced analyses
23.4	31.0	26.8	34.0	Mn
15.7	20.0	23.8	13.6	Fe
0.47	0.75	0.88	0.51	Co
1.47	0.73	0.59	1.72	Ni
1.51	0.52	0.34	1.42	Cu
0.54	0.26	0.29	0.17	Pb

TABLE

Station	MP 3	DWBD 1	Alb. 2	DWHG 4
Latitude	N 20°51′	N 21°27′	N 28°23′	N 24°22′
Longitude	W 127°16′	W 126°43′	W 126°57′	W 125°00′
Depth (m)	—	4,300	4.340	4,330
Asso. sed.	R. C.	R. C.	R. C.	R. C.
Sampler	—	Dredge	Trawl	Core
Nod. dia. (cm)	1.1	4×4×4	1–15	1×2×2
Sp. G.	—	2.33	2.47	—
Sect. sampled	Out. 2 cm	x-sect.	Whole nod.	Whole nod.

Chemical analyses (Weight percentages)

Al_2O_3	6.8	3.6	11.1	3.8
SiO_2	17.0	40.3	31.6	12.2
P	—	0.22	0.128	—
K	0.81	0.90	1.83	0.34
Ca	1.36	1.00	0.75	1.3
Ti	0.48	0.58	0.43	0.5
Mn	21.2	9.7	10.4	18.5
Fe	9.2	11.5	10.6	9.5
Co	0.36	0.30	0.19	0.24
Ni	1.10	0.13	0.67	1.21
Cu	0.76	0.19	0.44	0.49
Zn	0.095	0.043	0.050	0.03
Sr	0.084	0.066	0.062	0.056
Mo	0.049	0.028	0.032	0.039
Ba	0.55	0.42	0.42	0.18
Pb	0.15	0.12	0.13	0.051
H_2O	18.9	12.5	14.3	—
L.O.I.	—	—	—	28.5
Reference				GOLDBERG and MENARD (pers. comm., 1960)

Reduced analyses

Mn	37.0	22.3	24.2	30.1
Fe	16.1	26.4	24.6	15.4
Co	0.63	0.69	0.44	0.38
Ni	1.92	0.30	1.56	1.97
Cu	1.33	0.43	1.02	0.80
Pb	0.26	0.28	0.30	0.08

XXX (continued)

Vit. 4221	Wig. 6	Vit. 4217	Vit. 4217	Station
N 29°58′	N 28°59′	N 29°57′	N 29°57′	Latitude
W 125°55′	W 125°41′	W 120°42′	W 120°42′	Longitude
4,325	4,000	4,017	4,098	Depth (m)
R. C.	R. C.	R. C.	R. C.	Sampler
—	Core	Trawl	Trawl	Asso. sed.
2×1.5×1.5	0.3	4×2×2	3–12	Nod. dia. (cm)
2.47	—	2.54	—	Sp. G.
0.5 nod.	Whole nod.	x-sect.	Out. 0.5 cm	Sect. sampled
			Chemical analyses (Weight percentages)	
15.0	6.7	7.9	6.50	Al_2O_3
36.0	18.6	21.6	19.63	SiO_2
—	—	—	—	P
2.41	0.90	0.83	0.90	K
0.74	1.11	1.34	1.91	Ca
0.32	1.20	0.49	0.45	Ti
8.3	14.3	16.7	15.2	Mn
7.0	13.6	11.2	14.0	Fe
0.15	0.37	0.15	0.13	Co
0.41	0.51	0.74	0.43	Ni
0.25	0.34	0.45	—	Cu
0.067	0.081	0.069	—	Zn
0.065	0.079	0.063	—	Sr
0.011	0.036	0.042	—	Mo
0.24	0.57	0.59	—	Ba
0.16	0.18	0.12	0.10	Pb
15.3	20.8	18.9	18.44	H_2O
—	—	—	—	L.O.I.
			SKORNYA-KOVA et al. (1962)	Reference
				Reduced analyses
24.6	26.5	32.4	27.4	Mn
20.8	25.2	21.4	25.2	Fe
0.44	0.69	0.29	0.23	Co
1.22	0.95	1.44	0.77	Ni
0.74	0.63	0.87	—	Cu
0.48	0.33	0.23	—	Pb

TABLE

Station	PAS 19121	Vit. 4265b	UNK MS	VS 78
Latitude	N 27°20'	N 24°58'	N 22°30'	N 29°03'
Longitude	W 116°10'	W 113°25'	W 113°08'	W 113°33'
Depth (m)	4,030	3,330	3,604	384–493
Asso. sed.	—	T. Sed.	R. C.	T. Sed.
Sampler	Core	Trawl	Dredge	Dredge
Nod. dia. (cm)	0.5×2×2	5–6	5×3×3	8×6×5
Sp. G.	—	—	2.54	2.45
Sect. sampled	Whole nod.	Out. 1 cm	x-sect.	x-sect.
Chemical analyses (Weight percentages)				
Al_2O_3	4.5	4.97	7.3	3.6
SiO_2	13.3	11.63	17.1	8.9
P	—	—	—	—
K	0.7	0.99	0.95	0.96
Ca	1.0	1.83	1.28	1.16
Ti	0.6	—	0.16	0.07
Mn	21.2	32.8	28.8	38.9
Fe	9.3	1.2	4.85	0.86
Co	0.27	0.00	0.026	0.010
Ni	1.25	0.18	0.63	0.045
Cu	0.70	—	0.42	0.010
Zn	0.04	—	0.026	0.023
Sr	0.051	—	0.062	0.10
Mo	0.048	—	0.065	0.022
Ba	0.20	—	0.43	0.37
Pb	0.65	—	0.022	0.025
H_2O	—	15.32	17.3	13.0
L.O.I.	22.0	—	—	—
Reference	GOLDBERG and MENARD (pers. comm., 1960)	SKORNYAKOVA et al. (1962)		
Reduced analyses				
Mn	32.0	48.0	49.5	52.2
Fe	14.1	1.7	8.3	1.2
Co	0.41	0.00	0.045	0.013
Ni	1.89	0.26	1.08	0.060
Cu	1.06	—	0.72	0.013
Pb	0.098	—	0.038	0.034

XXX (continued)

VS BII-35	UNK BH2	Vit. 3729	Vit. 3899	Station
N 22°18'	N 13°37'	N 15°32'	N 17°00'	Latitude
W 107°48'	E 126°27'	E 134°30'	E 141°43'	Longitude
3,000	5,180	3,590	4,620	Depth (m)
T. Sed.	R. C.	—	R. C.	Asso. sed.
Trawl	Tele. cable	—	Trawl	Sampler
5×2×1	90×90×120	—	1–2.5	Nod. dia. (cm)
2.48	2.66	—	—	Sp. G.
x-sect.	Out. 1 cm	—	—	Sect. sampled
			Chemical analyses (Weight percentages)	
8.2	5.7	4.09	8.54	Al_2O_3
29.2	14.0	13.58	26.61	SiO_2
—	0.164	—	0.21	P
1.52	0.38	0.66	—	K
0.87	1.37	1.58	2.19	Ca
0.08	0.87	0.74	0.94	Ti
24.8	14.9	14.2	5.05	Mn
1.36	18.40	19.5	19.0	Fe
0.017	0.29	—	0.08	Co
0.12	0.22	0.39	0.05	Ni
0.046	0.17	—	—	Cu
0.043	0.048	—	—	Zn
0.039	0.11	—	—	Sr
0.032	0.037	—	—	Mo
0.33	0.58	—	—	Ba
0.046	0.18	—	—	Pb
11.0	21.4	21.66	—	H_2O
—	—	—	20.17	L.O.I.
		Skornya- kova et al. (1962)	Skornya- kova et al. (1962)	Reference
			Reduced analyses	
48.0	25.3	23.4	11.3	Mn
2.6	31.2	32.2	41.5	Fe
0.033	0.49	—	0.18	Co
0.23	0.37	0.64	0.11	Ni
0.089	0.29	—	—	Cu
0.089	0.31	—	—	Pb

TABLE

Station	Vit. 3631	MP 43D	MP 37A	MP 33K
Latitude	N 19°55'	N 11°57'	N 17°04'	N 17°48'
Longitude	E 155°59'	E 164°59'	W 177°15'	W 174°22'
Depth (m)	5,643	1,500–2,100	2,010–1,830	1,810–2,290
Asso. sed.	R. C.	—	Coral D.	Coral D.
Sampler	Spoon	Dredge	Dredge	Dredge
Nod. dia. (cm)	1.5–6	5 cm Crust	2 cm Crust	1 cm Crust
Sp. G.	—	—	2.39	—
Sect. sampled	4 cm nod.	—	x-sect.	x-sect.
Chemical analyses (Weight percentages)				
Al_2O_3	5.23	0.9	4.2	3.4
SiO_2	13.31	3.6	8.8	10.6
P	0.17	—	0.138	0.082
K	—	0.2	0.43	0.39
Ca	1.79	1.6	8.48	6.85
Ti	1.06	1.1	1.00	0.89
Mn	17.2	19.5	13.0	14.4
Fe	14.4	11.5	10.7	14.05
Co	0.53	1.05	0.45	0.70
Ni	0.33	0.42	0.47	0.29
Cu	—	0.11	0.19	0.072
Zn	—	—	0.057	0.04
Sr	—	0.11	0.17	0.14
Mo	—	0.039	0.049	0.043
Ba	—	0.15	0.56	0.43
Pb	—	0.15	0.19	0.20
H_2O	—	—	22.5	19.4
L.O.I.	27.72	34.5	—	—
Reference	SKORNYA-KOVA et al. (1962)	GOLDBERG and MENARD (pers. comm., 1960)		
Reduced analyses				
Mn	32.0	28.7	26.9	26.4
Fe	26.8	16.9	22.2	25.8
Co	0.98	1.54	0.93	1.28
Ni	0.61	0.62	0.97	0.53
Cu	—	0.16	0.39	0.13
Pb	—	0.22	0.39	0.37

XXX (continued)

MP 32	MP 26A3	MP 25F2	Tet. 22	Station
N 18°20′	N 19°03′	N 19°07′	N 16°06′	Latitude
W 173°23′	W 171°00′	W 169°44′	W 165°45′	Longitude
3,860	1,372	1,740	2,400	Depth (m)
R. C.	Volc. R.	Volc. R.	—	Asso. sed.
Dredge	Dredge	Dredge	Dredge	Sampler
Crust	3×3×3	3×2×2	2×2.5×3	Nod. dia. (cm)
—	2.45	2.16	—	Sp. G.
Out. 0.5 cm	x-sect.	x-sect.	x-sect.	Sect. sampled

				Chemical analyses (Weight percentages)
4.3	1.9	1.9	3.2	Al_2O_3
15.1	5.7	7.0	12.8	SiO_2
—	0.074	0.031	—	P
0.52	0.38	0.40	—	K
1.51	2.24	2.07	2.01	Ca
1.10	1.13	1.13	0.84	Ti
13.1	22.7	20.5	16.0	Mn
14.6	13.3	14.5	17.2	Fe
0.42	0.95	0.95	0.73	Co
0.30	0.60	0.42	0.25	Ni
0.17	0.15	0.10	0.04	Cu
0.043	0.062	0.062	0.053	Zn
0.12	0.14	0.15	0.12	Sr
0.027	0.071	0.060	0.025	Mo
0.61	0.58	0.50	0.45	Ba
0.22	0.23	0.24	0.19	Pb
25.2	23.9	28.2	20.6	H_2O
—	—	—	—	L.O.I.
				Reference

				Reduced analyses
24.5	33.2	32.6	25.2	Mn
27.4	19.5	23.0	27.2	Fe
0.79	1.39	1.51	1.15	Co
0.56	0.88	0.67	0.40	Ni
0.32	0.22	0.17	0.06	Cu
0.41	0.34	0.38	0.08	Pb

TABLE

Station	Tet. 27A	Msn G	Chal. 264	Msn 150G
Latitude	N 13°05'	N 14°11'	N 14°19'	N 10°59'
Longitude	W 163°10'	W 161°08'	W 152°37'	W 142°37'
Depth (m)	5,413	5,652	5,494	4,978
Asso. sed.	R. C.	R. C.	R. C.	R. C.
Sampler	Dredge	Corer	Trawl	Core
Nod. dia. (cm)	1.2	0.5–1 × 3	—	1 × 0.8 × 1
Sp. G.	—	2.52	—	—
Sect. sampled	Whole nod.	Whole nod.	Whole nod.	Whole nod.
Chemical analyses (Weight percentages)				
Al_2O_3	7.6	6.1	3.2	7.8
SiO_2	18.9	12.0	28.2	21.2
P	—	—	—	—
K	—	0.64	—	—
Ca	1.64	1.49	1.54	1.23
Ti	0.53	0.75	—	0.36
Mn	18.5	23.3	18.4	17.0
Fe	10.0	9.2	16.2	6.4
Co	0.31	0.31	—	0.31
Ni	0.86	0.98	—	1.23
Cu	0.65	0.81	—	0.96
Zn	0.084	0.11	—	0.087
Sr	0.084	0.080	—	0.075
Mo	0.031	0.053	—	0.029
Ba	0.40	0.46	—	0.33
Pb	0.12	0.085	—	0.10
H_2O	11.2	18.4	8.9	11.5
L.O.I.	—	—		—
Reference			Murray and Renard (1891)	
Reduced analyses				
Mn	29.7	36.8	29.4	29.0
Fe	16.1	14.4	25.8	11.0
Co	0.50	0.49	—	0.53
Ni	1.38	1.55	—	2.10
Cu	1.05	1.28	—	1.64
Pb	0.13	0.13	—	0.15

XXX (continued)

Msn 153P	Msn 153P	Car.78	MP 5	Station
N 13°07′	N 13°07′	N 16°15′	N 14°22′	Latitude
W 138°56′	W 138°56′	W 137°06′	W 133°07′	Longitude
4,927	4,927	4,553	—	Depth (m)
R. C.	(26 cm down	R. C.	R. C.	Asso. sed.
Core	in the core)	Snapper	—	Sampler
1×1.2×1		1	0.5	Nod. dia. (cm)
—	—	—	—	Sp. G.
Whole nod.	x-sect.	Whole nod.	Out. 1 cm	Sect. sampled
				Chemical analyses (Weight percentages)
5.1	3.8	5.3	5.4	Al_2O_3
13.5	11.0	29.4	15.0	SiO_2
—	—	—	—	P
—	—	2.0	0.77	K
1.39	1.57	0.8	1.36	Ca
0.31	0.48	0.8	0.09	Ti
25.0	23.6	10.0	22.9	Mn
5.1	7.9	8.9	9.2	Fe
0.33	0.34	0.20	0.45	Co
1.50	0.98	0.46	1.05	Ni
1.31	0.86	0.40	0.78	Cu
0.12	0.10	0.04	0.11	Zn
0.063	0.08	0.07	0.079	Sr
0.038	0.047	0.015	0.037	Mo
0.36	0.59	0.095	0.51	Ba
0.09	0.10	0.11	0.12	Pb
13.5	17.4	—	19.3	H_2O
—	—	21.0	—	L.O.I.
		GOLDBERG and MENARD (pers. comm., 1960)		Reference
				Reduced analyses
36.8	34.8	19.9	38.0	Mn
7.5	11.6	17.7	15.3	Fe
0.49	0.50	0.40	0.75	Co
2.20	1.45	0.92	1.74	Ni
1.93	1.26	0.80	1.30	Cu
0.18	0.14	0.22	0.20	Pb

TABLE

Station	Chub. 2	Cap. 50B	Chub. 1	UNK RR	Trans. 14C
Latitude	N 16°00′	N 14°55′	N 19°00′	N 19°49′	N 19°46′
Longitude	W 125°01′	W 124°12′	W 121°53′	W 121°44′	W 114°44′
Depth (m)	4,354	4,270	4,138	4,320	3,438
Asso. sed.	R. C.	R. C.	R. C.	R. C.	R. C.
Sampler	Core	Core	Core	Dredge	Core
Nod. dia. (cm)	0.1	—	0.1	1–1.5	1×3×2
Sp. G.	—	—	—	2.80	2.43
Sect. sampled	—	x-sect.	Whole nod.	Whole nod.	x-sect.
Chemical analyses (Weight percentages)					
Al_2O_3	6.4	6.1	5.8	7.1	4.8
SiO_2	15.9	16.1	14.7	16.8	12.7
P	—	—	—	0.089	—
K	0.88	0.82	0.74	1.04	0.60
Ca	1.39	1.19	1.34	1.27	1.50
Ti	0.44	0.33	0.52	0.59	0.48
Mn	23.8	22.4	22.4	21.4	21.2
Fe	7.3	7.6	9.5	9.6	12.0
Co	0.27	0.39	0.40	0.36	0.22
Ni	1.22	1.15	1.16	1.09	0.93
Cu	1.05	1.25	0.87	0.76	0.61
Zn	0.14	0.11	0.10	0.086	0.083
Sr	0.078	0.074	0.090	0.092	0.10
Mo	0.049	0.052	0.052	0.046	0.047
Ba	0.52	0.42	0.70	0.57	0.46
Pb	0.082	0.11	0.14	0.12	0.098
H_2O	17.9	18.3	19.7	20.3	20.5
Reference					
Reduced analyses					
Mn	39.8	37.7	37.5	38.3	34.2
Fe	12.2	12.8	15.9	13.1	19.4
Co	0.45	0.66	0.67	0.65	0.36
Ni	2.04	1.94	1.94	1.95	1.50
Cu	1.76	2.10	1.46	1.36	0.99
Pb	0.14	0.19	0.23	0.22	0.16

XXX (continued)

Trans. 14D	SW 48	Acap. 11	Acap. 10	Station
N 19°20′	N 11°25′	N 10°53′	N 11°38′	Latitude
W 114°12′	W 113°48′	W 105°07′	W 103°48′	Longitude
3,480	4,085	3,275	3,500	Depth (m)
R. C.	R. C.	—	—	Asso. sed.
Core	Core	Core	Core	Sampler
3×3×1	—	1	0.1	Nod. dia. (cm)
2.70	—	—	—	Sp. G.
x-sect.	—	Whole nod.	—	Sect. sampled

				Chemical analyses (Weight percentages)
5.0	3.6	12.5	6.9	Al_2O_3
13.1	9.0	28.5	23.8	SiO_2
—	—	—	—	P
0.65	0.48	0.65	0.65	K
1.44	1.79	2.97	12.6	Ca
0.36	0.66	0.14	0.22	Ti
22.6	23.2	3.4	1.7	Mn
10.2	11.5	15.5	6.3	Fe
0.23	0.21	0.03	0.013	Co
1.09	1.01	0.036	0.045	Ni
0.71	0.66	0.078	0.076	Cu
0.11	0.086	0.019	0.057	Zn
0.084	0.11	0.17	0.12	Sr
0.041	0.061	0.006	0.018	Mo
0.61	0.56	0.70	0.71	Ba
0.053	0.12	0.090	0.082	Pb
20.4	23.2	14.0	20.0	H_2O
				Reference

				Reduced analyses
36.8	36.2	7.6	7.8	Mn
16.1	17.9	34.5	28.9	Fe
0.37	0.33	0.067	0.06	Co
1.77	1.58	0.08	0.21	Ni
1.15	1.03	0.17	0.35	Cu
0.086	0.19	0.20	0.38	Pb

TABLE

Station	Msn K	Msn J	Msn 139D	Alb. 13
Latitude	N 6°03′	N 7°47′	S 0°45′	N 9°57′
Longitude	W 170°00′	W 168°00′	W 147°36′	W 137°47′
Depth (m)	5,400	4,994	3,340	4,930
Asso. sed.	R. C.	R. C.	Calc. O.	R. C.
Sampler	Core	Core	Dredge	Trawl
Nod. dia. (cm)	1.5×1.5	2×2.5×2.5	1–5	10–16
Sp. G.	2.70	2.28	2.07	2.53
Sect. sampled	x-sect.	x-sect.	x-sect.	x-sect.
Chemical analyses (Weight percentages)				
Al_2O_3	6.0	3.3	1.3	5.7
SiO_2	11.3	9.4	7.6	13.0
P	—	—	—	0.052
K	0.54	0.41	0.35	0.79
Ca	1.46	1.73	1.98	1.47
Ti	0.25	1.23	1.01	0.44
Mn	29.0	20.2	18.6	29.8
Fe	5.25	13.8	17.3	4.8
Co	0.16	0.39	0.44	0.20
Ni	1.54	0.60	0.32	1.36
Cu	1.90	0.43	0.11	1.20
Zn	0.16	0.069	0.060	0.12
Sr	0.057	0.12	0.14	0.070
Mo	0.052	0.042	0.054	0.054
Ba	0.23	0.59	0.49	0.61
Pb	0.053	0.11	0.12	0.055
H_2O	17.7	24.3	28.0	16.2
L.O.I.	—	—	—	22.0
Reference				
Reduced analyses				
Mn	44.6	33.7	29.5	45.3
Fe	8.1	21.9	27.4	7.3
Co	0.25	0.62	0.70	0.30
Ni	2.37	0.95	0.51	2.07
Cu	2.92	0.68	0.17	1.83
Pb	0.082	0.17	0.19	0.084

XXX (continued)

Chub. 19	Chub. 39	DWBG 147B	Alb. 4622	Station
N 7°41'	N 8°09'	N 1°27'	N 6°21'	Latitude
W 125°37'	W 125°20'	W 116°13'	W 81°44'	Longitude
4,416	4,360	4,000	1,061	Depth (m)
Rad. O.	Rad. O.	Calc. O.	T. Sed.	Asso. sed.
Core	Core	Core	Dredge	Sampler
0.6	1.0	1×1.5×1.5	3×4×1.5	Nod. dia. (cm)
—	—	—	2 34	Sp. G.
0.5 Nod.	x-sect.	Whole nod.	x-sect.	Sect. sampled
			Chemical analyses (Weight percentages)	
8.7	4.2	1.7	4.3	Al_2O_3
30.7	12.5	7.5	15.5	SiO_2
—	—	—	0.253	P
2.05	0.72	—	0.40	K
0.72	1.51	1.8	1.70	Ca
0.18	0.18	0.3	0.47	Ti
9.3	28.1	18.8	15.1	Mn
9.2	6.3	12.6	17.7	Fe
0.09	0.18	0.05	0.36	Co
0.40	1.16	0.76	0.25	Ni
0.55	1.36	0.47	0.05	Cu
0.093	0.15	0.04	0.043	Zn
0.055	0.067	0.10	0.15	Sr
0.043	0.047	0.03	0.037	Mo
0.18	0.45	0.18	0.45	Ba
0.08	0.058	0.045	0.082	Pb
16.5	19.0	—	20.1	H_2O
—	—	31.0	26.0	L.O.I.
		GOLDBERG and MENARD (pers. comm., 1960)		*Reference*
				Reduced analyses
21.1	43.7	28.6	25.2	Mn
20.9	9.8	19.2	29.4	Fe
0.22	0.28	0.076	0.60	Co
0.91	1.80	1.16	0.42	Ni
1.25	2.12	0.72	0.083	Cu
0.18	0.09	0.068	0.14	Pb

TABLE

Station	DWBD 2	Chub. 5	Chub. 3	Vit. 3996a
Latitude	N 10°26′	N 13°03′	N 15°00′	N 4°57′
Longitude	W 130°38′	W 125°29′	W 125°26′	E 135°30′
Depth (m)	4,890	4,440	4,380	4,580
Asso. sed.	Rad. O.	R. C.	R. C.	G. O.
Sampler	Dredge	Core	Core	Trawl
Nod. dia. (cm)	4×3×1	—	1	5–9
Sp. G.	2.56	—	—	—
Sect. sampled	0.5 nod.	Whole nod.	x-sect.	Out. layer
Chemical analyses (Weight percentages)				
Al_2O_3	6.3	7.5	5.3	4.96
SiO_2	15.7	20.8	12.6	15.31
P	0.052	—	—	0.14
K	0.86	0.98	0.69	—
Ca	1.46	1.18	1.42	2.01
Ti	0.41	0.38	0.74	0.47
Mn	22.7	22.2	22.2	19.1
Fe	7.6	6.3	9.7	12.0
Co	0.26	0.32	0.38	0.20
Ni	1.25	1.06	1.00	0.65
Cu	1.21	1.06	0.82	—
Zn	0.11	0.088	0.081	—
Sr	0.073	0.056	0.096	—
Mo	0.059	0.041	0.054	—
Ba	0.59	0.74	0.67	—
Pb	0.075	0.078	0.17	—
H_2O	19.0	17.4	20.4	—
L.O.I.	—	—	—	25.97
Reference				Skornya-kova et al. (1962)
Reduced analyses				
Mn	38.5	41.0	36.0	35.4
Fe	12.9	11.6	15.7	22.4
Co	0.44	0.59	0.62	0.37
Ni	2.12	1.95	1.62	1.20
Cu	2.05	1.95	1.33	—
Pb	0.14	0.14	0.28	—

XXX (continued)

Msn 148G	Vit. 3802	Msn 128G	Msn 128G	Station
N 9°06′	S 3°17′	S 13°53′	S 13°53′	Latitude
W 145°18′	W 172°52′	W 150°35′	W 150°35′	Longitude
5,400	5,329	3,623	3,625	Depth (m)
R. C.	R. C.	Coral S.	(Nodule	Asso. sed.
Core	Trawl	Core	from 14 cm	Sampler
1.5×2.5×2.7	2.5	3×3×2	down in	Nod. dia. (cm)
—	—	—	core)	Sp. G.
x-sect.	Whole nod.	x-sect.	x-sect.	Sect. sampled
			Chemical analyses (Weight percentages)	
5.1	6.31	3.2	4.2	Al_2O_3
13.6	14.71	8.9	11.7	SiO_2
—	0.19	—	—	P
—	—	—	—	K
1.45	2.15	1.96	1.80	Ca
0.31	0.46	0.68	0.78	Ti
26.2	24.4	15.7	14.3	Mn
5.3	7.2	17.5	17.4	Fe
0.26	0.24	0.43	0.43	Co
1.52	0.75	0.31	0.31	Ni
1.27	—	0.21	0.24	Cu
0.12	—	0.053	0.060	Zn
0.060	—	0.12	0.11	Sr
0.043	—	0.028	0.022	Mo
0.38	—	0.41	0.39	Ba
0.09	—	0.13	0.13	Pb
14.9	—	17.4	19.6	H_2O
—	21.77	—	—	L.O.I.
	Skornya-kova et al. (1962)			Reference
				Reduced analyses
39.5	42.7	22.3	22.2	Mn
8.0	12.6	24.8	27.0	Fe
0.39	0.42	0.61	0.67	Co
2.28	1.31	0.44	0.48	Ni
1.91	—	0.30	0.37	Cu
0.18	—	0.075	0.09	Pb

TABLE

Station	Msn S	Msn Q	Chal. 274	Alb. 4721
Latitude	S 9°00′	S 7°03′	S 7°25′	S 8°08′
Longitude	E 171°28′	E 174°12′	W 152°15′	W 104°11′
Depth (m)	5,000	5,378	5,040	3,820
Asso. sed.	R. C.	R. C.	Rad. O.	Calc. O.
Sampler	Dredge	Core	Trawl	Trawl
Nod. dia. (cm)	0.5–4	1.5	$5 \times 10 \times 10$	2–8
Sp. G.	2.14	2.54	—	2.58
Sect. sampled	Whole nod.	0.5 nod.	Whole nod.	Whole nod.
Chemical analyses (Weight percentages)				
Al_2O_3	7.8	7.6	0.6	2.5
SiO_2	15.6	13.8	11.4	10.0
P	—	—	0.07	0.113
K	0.52	0.42	—	0.49
Ca	1.84	1.45	2.11	1.39
Ti	0.42	1.02	—	0.14
Mn	19.5	15.7	35.3	23.0
Fe	11.1	15.5	7.4	14.3
Co	0.13	0.26	—	0.55
Ni	0.63	0.45	—	1.00
Cu	0.71	0.45	0.67	0.70
Zn	0.12	0.060	—	0.11
Sr	0.083	0.10	—	0.082
Mo	0.021	0.037	—	0.032
Ba	0.22	0.41	—	0.50
Pb	0.080	0.13	—	0.050
H_2O	19.4	21.8	11.4	17.8
L.O.I.	—	—	—	26.5
Reference			Murray and Renard (1891)	
Reduced analyses				
Mn	34.2	27.6	45.1	33.0
Fe	19.4	27.3	9.5	20.5
Co	0.23	0.46	—	0.079
Ni	1.10	0.79	—	1.43
Cu	1.24	0.79	0.86	1.00
Pb	0.14	0.23	—	0.072

XXX (continued)

Alb. 4711	Alb. 4660	Alb. 4658	Alb. 4656	Station
S 7°48'	S 9°56'	S 8°30'	S 6°55'	Latitude
W 94°06'	W 87°30'	W 85°36'	W 83°34'	Longitude
4,100	4,440	4,330	4,060	Depth (m)
Calc. O.	R. C.	Rad. O.	Gr. Mud	Asso. sed.
Trawl	Trawl	Trawl	Trawl	Sampler
3–15	4–18	8–12	4×8×10	Nod. dia. (cm)
2.69	2.48	2.97	2.52	Sp. G.
x-sect.	x-sect.	x-sect.	x-sect.	Sect. sampled
				Chemical analyses (Weight percentages)
3.4	4.1	2.9	8.4	Al_2O_3
7.9	9.5	6.3	44.0	SiO_2
0.101	0.041	0.026	0.020	P
0.62	0.98	0.89	1.54	K
1.74	1.07	1.24	0.69	Ca
0.20	0.10	0.06	0.11	Ti
36.5	39.1	42.3	9.6	Mn
3.68	1.98	0.83	7.71	Fe
0.050	0.030	0.006	0.027	Co
1.10	0.57	0.14	0.11	Ni
0.57	0.30	0.13	0.19	Cu
0.13	0.074	0.043	0.067	Zn
0.055	0.048	0.039	0.043	Sr
0.038	0.041	0.045	0.022	Mo
0.59	0.73	0.26	0.42	Ba
0.011	0.004	0.022	0.078	Pb
16.2	17.9	13.9	11.9	H_2O
21.5	23.5	—	18.5	L.O.I.
				Reference
				Reduced analyses
50.4	57.1	55.0	26.9	Mn
5.1	2.9	1.1	21.6	Fe
0.069	0.043	0.008	0.078	Co
1.52	0.83	0.18	0.31	Ni
0.79	0.43	0.17	0.53	Cu
0.015	0.006	0.029	0.22	Pb

TABLE

Station	Chal. 276	Alb. 173	DWBD 4	DWBD 16	Alb. 31
Latitude	S 13°28'	S 18°55'	S 17°	S 16°29'	S 12°20'
Longitude	W 149°30'	W 146°23'	W 146°	W 145°33'	W 144°15'
Depth (m)	4,300	4,460	1,700	1,270	4,840
Asso. sed.	R. C.	R. C.	Coral D.	Coral D.	R. C.
Sampler	Trawl	Trawl	Dredge	Dredge	Trawl
Nod. dia. (cm)	2	4×6–10	3×4×7	1–5	1
Sp. G.	2.57	2.40	2.31	2.38	—
Sect. sampled	x-sect.	x-sect.	x-sect.	Whole nod.	Whole nod.

Chemical analyses (Weight percentages)

Al_2O_3	5.3	6.4	0.7	1.3	7.0
SiO_2	12.5	13.4	1.5	2.8	14.6
P	0.127	0.213	0.122	0.171	—
K	0.58	0.50	0.34	0.32	1.2
Ca	1.87	2.08	3.14	2.92	1.2
Ti	0.88	1.20	1.11	1.18	0.5
Mn	21.6	15.0	23.2	22.4	19.3
Fe	12.0	16.1	12.6	13.8	7.9
Co	0.35	0.50	1.53	1.10	0.15
Ni	0.77	0.23	0.58	0.58	0.81
Cu	0.35	0.17	0.095	0.17	0.61
Zn	0.071	0.057	0.062	0.067	0.03
Sr	0.11	0.12	0.15	0.16	0.05
Mo	0.043	0.031	0.050	0.056	0.034
Ba	0.25	0.54	0.66	0.51	0.11
Pb	0.13	0.12	0.28	0.25	0.032
H_2O	23.6	17.7	24.5	27.5	—
L.O.I.	—	22.5	35.0	36.0	19.5
Reference					GOLDBERG and MENARD (pers. comm., 1960)

Reduced analyses

Mn	36.9	24.0	31.7	32.8	29.8
Fe	20.5	25.8	17.2	20.2	12.2
Co	0.60	0.80	2.09	1.61	0.23
Ni	1.32	0.37	0.79	0.85	1.25
Cu	0.60	0.27	0.13	0.25	0.94
Pb	0.22	0.19	0.38	0.37	0.049

XXX (continued)

Car. 46	DWHD 15	DWBG 19	DWBG 17	Station
S 17°36'	S 15°23'	S 14°59'	S 12°51'	Latitude
W 141°55'	W 136°18'	W 136°02'	W 135'13'	Longitude
2,132	4,480	4,465	4,318	Depth (m)
Calc. O.	Calc. O.	R. C.	R. C.	Asso. sed.
—	Dredge	Core	Core	Sampler
1	0.5×3×3	0.5–1	0.5	Nod. dia. (cm)
—	2.35	2.38	—	Sp. G.
Whole nod.	x-sect.	x-sect.	Whole nod.	Sect. sampled
			Chemical analyses (Weight percentages)	
2.1	8.6	13.9	4.3	Al_2O_3
6.9	15.8	30.4	10.1	SiO_2
—	0.223	—	—	P
0.35	1.44	1.95	0.4	K
2.0	1.39	0.90	1.6	Ca
1.0	0.29	0.64	0.3	Ti
16.0	20.3	12.0	23.9	Mn
16.2	8.1	9.8	7.0	Fe
0.49	0.12	0.12	0.085	Co
0.34	1.17	0.77	1.46	Ni
0.09	1.07	0.53	0.92	Cu
0.035	0.12	0.083	0.046	Zn
0.077	0.067	0.040	0.032	Sr
0.035	0.041	0.018	0.03	Mo
0.091	0.24	0.22	0.19	Ba
0.07	0.060	0.038	0.028	Pb
—	20.3	14.2	—	H_2O
30.0	22.5	22.0	23.0	L.O.I.
GOLDBERG and MENARD (pers. comm., 1960)			GOLDBERG and MENARD (pers. comm., 1960)	Reference
				Reduced analyses
23.2	36.7	28.9	34.4	Mn
23.5	14.6	23.6	10.1	Fe
0.71	0.22	0.29	0.12	Co
0.49	2.12	1.86	2.10	Ni
0.13	1.94	1.28	1.32	Cu
0.10	0.11	0.092	0.040	Pb

TABLE

Station	Alb. 4701	Alb. 4662	Alb. 4681	Alb. 4676
Latitude	S 19°12′	S 11°14′	S 18°47′	S 14°29′
Longitude	W 102°24′	W 89°35′	W 89°26′	W 81°24′
Depth (m)	4,150	4,460	4,390	4,970
Asso. sed.	R. C.	Rad. O.	R. C.	Calc. O.
Sampler	Trawl	Trawl	Trawl	Trawl
Nod. dia. (cm)	0.5×2×2	1–10	2–3	5–6
Sp. G.	—	2.41	2.50	2.46
Sect. sampled	Whole nod.	x-sect.	x-sect.	x-sect.

Chemical analyses (Weight percentages)

Al_2O_3	3.8	4.6	5.0	7.2
SiO_2	17.5	12.2	10.0	17.1
P	—	0.161	0.040	0.377
K	1.4	0.71	0.65	0.73
Ca	1.0	1.40	1.72	1.77
Ti	0.17	0.35	0.34	0.21
Mn	17.2	25.1	26.6	23.4
Fe	11.6	8.88	8.80	7.75
Co	0.16	0.13	0.24	0.053
Ni	1.52	1.11	1.26	1.25
Cu	0.59	0.78	0.76	0.80
Zn	—	0.095	0.14	0.11
Sr	0.041	0.092	0.077	0.064
Mo	0.043	0.033	0.047	0.058
Ba	0.22	0.59	0.41	0.27
Pb	0.052	0.060	0.082	0.049
H_2O	—	16.3	18.6	18.3
L.O.I.	20.0	23.5	26.5	22.0
Reference	GOLDBERG and MENARD (pers. comm., 1960)			

Reduced analyses

Mn	26.6	37.5	40.0	40.8
Fe	17.9	13.3	13.3	1.35
Co	0.25	0.19	0.36	0.092
Ni	0.35	1.66	1.90	2.18
Cu	0.91	1.17	1.14	1.40
Pb	0.080	0.090	0.12	0.085

XXX (continued)

Chal. 281	DWBG 37	Alb. 4685	DWHD 72	Station
S 22°21'	S 29°09'	S 21°36'	S 25°31'	Latitude
W 150°17'	W 143°01'	W 96°56'	W 85°14'	Longitude
4,365	4,120	4,040	920	Depth (m)
R. C.	R. C.	R. C.	Coral D.	Asso. sed.
Trawl	Core	Trawl	Dredge	Sampler
1	2×4×4	0.5–2	2–3	Nod. dia. (cm)
—	—	2.67	3.07	Sp. G.
Whole nod.	x-sect.	Whole nod.	x-sect.	Sect. sampled

				Chemical analyses (Weight percentages)
3.2	3.8	8.3	1.3	Al_2O_3
22.7	11.6	19.7	0.70	SiO_2
—	—	0.132	0.430	P
—	0.5	1.23	1.20	K
1.81	3.1	1.40	1.73	Ca
—	0.63	0.72	0.21	Ti
14.1	12.7	15.5	42.3	Mn
21.2	15.5	10.6	2.47	Fe
—	0.26	0.18	0.17	Co
—	0.33	1.09	0.26	Ni
—	0.16	0.62	0.15	Cu
—	—	0.10	0.052	Zn
—	0.06	0.066	0.090	Sr
—	0.024	0.033	0.043	Mo
—	0.08	0.32	0.79	Ba
—	0.08	0.082	0.062	Pb
16.0	—	19.0	12.8	H_2O
—	28.5	—	—	L.O.I.
Murray and Renard (1891)	Goldberg and Menard (pers. comm., 1960)			Reference

				Reduced analyses
23.0	19.2	28.9	49.7	Mn
34.6	23.5	19.8	2.9	Fe
—	0.39	0.33	0.20	Co
—	0.50	2.03	0.31	Ni
—	0.24	1.16	0.18	Cu
—	0.12	0.15	0.073	Pb

TABLE

Station	DWBG 40	Msn 121G	Msn 121G	Msn 125G
Latitude	S 31°13′	S 29°35′	S 29°35′	S 26°01′
Longitude	W 141°12′	W 158°58′	W 158°58′	W 155°59′
Depth (m)	4,280	5,252	5,252	5,038
Asso. sed.	Calc. O.	R. C.	(Nodule	R. C.
Sampler	Core	Core	48 cm down	Core
Nod. dia (cm)	—	$1.5 \times 1.5 \times 2.4$	in core)	$3 \times 3.5 \times 4$
Sp. G.	—	—	—	—
Sect. sampled	—	0.5 nod.	x-sect.	x-sect.

Chemical analyses (Weight percentages)

Al_2O_3	2.6	6.4	5.9	4.7
SiO_2	6.9	14.9	13.5	12.6
P	—	—	—	—
K	—	—	—	—
Ca	1.5	0.96	1.19	1.37
Ti	0.9	0.86	0.89	0.98
Mn	14.3	12.8	13.4	14.1
Fe	18.2	17.8	18.4	18.2
Co	0.29	0.49	0.54	0.73
Ni	0.24	0.29	0.33	0.32
Cu	0.12	0.20	0.23	0.14
Zn	—	0.06	0.056	0.052
Sr	0.10	0.081	0.089	0.093
Mo	0.035	0.023	0.021	0.020
Ba	0.12	0.56	0.68	0.72
Pb	0.069	0.16	0.17	0.18
H_2O	—	16.5	17.1	16.4
L.O.I.	31.5	—	—	—
Reference	GOLDBERG and MENARD (pers. comm., 1960)			

Reduced analyses

Mn	21.7	20.6	21.1	18.5
Fe	27.6	28.6	29.0	23.9
Co	0.44	0.79	0.85	0.96
Ni	0.36	0.47	0.52	0.42
Cu	0.18	0.32	0.36	0.18
Pb	0.10	0.10	0.09	0.068

XXX (continued)

Msn 126G	Msn 126G	Msn 116P	Msn 116P	Station
S 24°41′	S 24°41′	S 35°50′	S 35°50′	Latitude
W 154°45′	W 154°45′	W 163°01′	W 163°01′	Longitude
4,542	4,542	4,950	4,950	Depth (m)
R. C.	(Nodule	R. C.	R. C.	Asso. sed.
Core	29 cm	Core	(Nodule	Sampler
1×3×3	below	—	98 cm	Nod. dia. (cm)
—	surface)	—	below	Sp. G.
x-sect.	Out. 5 cm	—	surface)	Sect. sampled
				Chemical analyses (Weight percentages)
4.1	4.1	11.3	5.0	Al_2O_3
9.5	10.9	25.4	16.0	SiO_2
—	—	—	—	P
—	—	—	—	K
1.73	1.69	0.94	1.34	Ca
0.88	0.97	0.24	0.68	Ti
15.7	14.0	17.8	14.1	Mn
17.0	18.9	5.2	17.1	Fe
0.57	0.58	0.20	0.35	Co
0.41	0.27	1.08	0.30	Ni
0.26	0.31	0.79	0.13	Cu
0.063	0.059	0.12	0.05	Zn
0.10	0.11	0.055	0.10	Sr
0.025	0.021	0.019	0.022	Mo
0.65	0.62	0.22	0.60	Ba
0.15	0.16	0.10	0.19	Pb
18.7	17.4	10.2	18.3	H_2O
—	—	—	—	L.O.I.
				Reference
				Reduced analyses
23.2	20.7	33.5	23.2	Mn
25.1	28.0	9.8	28.2	Fe
0.86	0.86	0.38	0.58	Co
0.60	0.40	2.04	0.50	Ni
0.37	0.46	1.49	0.21	Cu
0.09	0.087	0.22	0.08	Pb

TABLE

Station	DWBG 43	Chal. 285	DWBG 47	DWBG 47
Latitude	S 34°01′	S 32°36′	S 36°33′	S 36°33′
Longitude	W 138°55′	W 137°43′	W 137°24′	W 137°24′
Depth (m)	4,721	4,350	4,700	4,700
Asso. sed.	R. C.	R. C.	R. C.	R. C.
Sampler	Core	Trawl	Core	(Nodule
Nod. dia. (cm)	$0.5 \times 2 \times 2$	3	$2 \times 1.5 \times 2$	72 cm
Sp. G.	—	—	—	below
Sect. sampled	x-sect.	Whole nod.	0.5 nod.	surface)
Chemical analyses (Weight percentages)				
Al_2O_3	5.4	5.5	6.2	5.6
SiO_2	16.1	13.4	18.8	15.5
P	—	0.057	—	—
K	0.3	0.21	—	—
Ca	2.2	1.65	1.44	1.18
Ti	1.4	0.078	0.57	0.48
Mn	19.6	16.7	14.1	18.1
Fe	21.7	10.1	12.4	10.5
Co	0.45	0.22	0.36	0.34
Ni	0.50	0.77	0.64	1.10
Cu	0.21	0.30	0.29	0.50
Zn	0.04	0.08	0.067	0.086
Sr	0.10	0.017	0.088	0.079
Mo	0.037	0.067	0.023	0.028
Ba	0.13	0.11	0.48	0.54
Pb	0.11	0.047	0.14	0.14
H_2O	—	29.95	16.7	15.2
L.O.I.	29.5	—	—	—
Reference	GOLDBERG and MENARD (pers. comm., 1960)	MURRAY and RENARD (1891)		
Reduced analyses				
Mn	35.7	30.6	24.2	28.4
Fe	39.6	18.5	21.5	16.5
Co	0.82	0.40	0.62	0.53
Ni	0.91	1.41	1.10	1.58
Cu	0.38	0.55	0.50	0.84
Pb	0.20	0.086	0.11	0.13

XXX (continued)

DWBG 46	DWBG 48	DWBG 48	Chal. 286	Station
S 36°23′	S 37°05′	S 37°05′	S 33°29′	Latitude
W 137°15′	W 137°00′	W 137°00′	W 133°22′	Longitude
4,680	4,940	4,940	4,270	Depth (m)
R. C.	R. C.	R. C.	R. C.	Asso. sed.
Core	Core	(Nodule	Trawl	Sampler
2.5×2.5×2.5	1.5×1.5×3	48 cm	0.5	Nod. dia. (cm)
—	—	below	—	Sp. G.
x-sect.	Whole nod.	surface)	Whole nod.	Sect. sampled

				Chemical analyses (Weight percentages)
4.2	4.9	3.5	2.8	Al_2O_3
12.2	12.3	8.9	20.0	SiO_2
—	—	—	0.14	P
—	—	—	—	K
1.50	1.35	1.81	2.3	Ca
0.61	0.48	0.31	—	Ti
18.7	18.9	20.2	24.1	Mn
13.0	11.2	12.6	12.5	Fe
0.38	0.38	0.16	—	Co
0.69	0.91	0.82	—	Ni
0.33	0.43	0.35	—	Cu
0.069	0.079	0.088	—	Zn
0.097	0.084	0.093	—	Sr
0.029	0.033	0.038	—	Mo
0.68	0.58	0.34	—	Ba
0.16	0.15	0.10	—	Pb
17.2	18.7	19.1	15.5	H_2O
—	—	—	—	L.O.I.
			Murray and Renard (1891)	Reference

				Reduced analyses
28.2	29.5	29.5	45.0	Mn
19.6	17.5	18.4	23.4	Fe
0.57	0.59	0.23	—	Co
1.04	1.42	1.20	—	Ni
0.50	0.67	0.51	—	Cu
0.10	0.12	0.13	—	Pb

TABLE

Station	Chal. 289	Chal. 293	Chal. 297	DWHD 55
Latitude	S 39°41′	S 39°04′	S 37°29′	S 37°04′
Longitude	W 131°23′	W 105°05′	W 83°07′	W 81°05′
Depth (m)	4,665	3,705	3,245	4,000
Asso. sed.	R. C.	Calc. O.	Calc. O.	Calc. O.
Sampler	Trawl	Trawl	Trawl	Dredge
Nod. dia. (cm)	6	2	2×2×2	2–3
Sp. G.	2.54	—	—	—
Sect. sampled	Whole nod.	Whole nod.	x-sect.	—
Chemical analyses (Weight percentages)				
Al_2O_3	6.2	3.2	6.4	5.3
SiO_2	11.8	16.7	14.3	13.1
P	0.08	0.14	—	—
K	0.64	—	—	0.3
Ca	1.97	2.73	1.40	1.2
Ti	0.88	—	0.20	0.22
Mn	20.7	23.8	17.2	19.1
Fe	12.0	14.6	12.7	7.1
Co	0.31	—	0.12	0.08
Ni	0.82	—	0.78	1.2
Cu	0.41	—	0.45	0.52
Zn	0.083	—	0.099	0.03
Sr	0.10	—	0.080	0.037
Mo	0.038	—	0.025	0.046
Ba	0.44	—	0.26	0.065
Pb	0.11	—	0.10	0.037
H_2O	20.8	11.2	14.9	—
L.O.I.	—	—	—	28.0
Reference		Murray and Renard (1891)		Goldberg and Menard (pers. comm., 1960)
Reduced analyses				
Mn	33.8	33.4	26.8	32.2
Fe	19.6	20.5	19.7	12.0
Co	0.51	—	0.19	0.13
Ni	1.34	—	1.21	2.02
Cu	0.67	—	0.70	0.88
Pb	0.18	—	0.15	0.062

XXX (continued)

Chal. 299	DWBG 52	DWBG 52	DWBG 54	Station
S 33°31'	S 40°36'	S 40°36'	S 41°24'	Latitude
W 74°43'	W 132°49'	W 132°49'	W 129°06'	Longitude
3,950	5,120	5,120	4,880	Depth (m)
Blue Mud	R. C.	(Nodule	R. C.	Asso. sed.
Trawl	Core	39 cm	Core	Sampler
3–4	1×1×1.5	below	2–3	Nod. dia. (cm)
—	—	surface)	2.41	Sp. G.
—	Whole nod.	Whole nod.	x-sect.	Sect. sampled
				Chemical analyses (Weight percentages)
4.7	5.2	4.5	6.3	Al_2O_3
12.2	14.5	12.7	14.8	SiO_2
—	—	—	—	P
0.6	—	—	0.62	K
1.3	1.21	1.31	1.51	Ca
0.14	0.57	0.47	1.04	Ti
29.0	18.6	21.6	19.6	Mn
2.5	10.3	8.1	11.7	Fe
0.008	0.43	0.40	0.39	Co
0.15	0.89	0.90	0.80	Ni
0.15	0.47	0.88	0.37	Cu
0.07	0.077	0.083	0.083	Zn
0.02	0.080	0.073	0.10	Sr
0.047	0.035	0.041	0.034	Mo
0.20	0.66	0.69	0.80	Ba
0.016	0.16	0.14	0.20	Pb
11.8	14.9	16.0	20.3	H_2O
18.0	—	—	—	L.O.I.
GOLDBERG				Reference
and MENARD				
(pers. comm., 1960)				
				Reduced analyses
40.6	28.5	32.4	33.5	Mn
3.5	15.8	12.1	20.0	Fe
0.011	0.66	0.60	0.067	Co
0.21	1.36	1.35	1.37	Ni
0.21	0.72	1.32	0.63	Cu
0.022	0.25	0.12	0.34	Pb

TABLE

Station	DWBG 56	DWBG 58	DWBG 7	DWHD 47
Latitude	S 42°16′	S 43°07′	S 46°44′	S 41°59′
Longitude	W 125°50′	W 125°23′	W 123°01′	W 102°01′
Depth (m)	4,560	4,640	4,100	4,200
Asso. sed.	R. C.	R. C.	Calc. O.	Calc. O.
Sampler	Core	Core	Dredge	Dredge
Nod. dia. (cm)	1.0	1.0	1	2–6
Sp. G.	—	—	2.41	2.42
Sect. sampled	Whole nod.	Whole nod.	x-sect.	x-sect.
Chemical analyses (Weight percentages)				
Al_2O_3	3.9	5.7	7.6	5.7
SiO_2	10.3	21.3	20.7	11.1
P	—	—	0.080	0.097
K	—	—	1.12	0.67
Ca	1.53	1.48	1.47	1.57
Ti	0.54	0.31	0.69	0.37
Mn	22.6	20.1	16.2	24.5
Fe	9.1	6.4	9.6	9.6
Co	0.40	0.25	0.23	0.13
Ni	1.10	1.14	0.86	1.02
Cu	0.45	0.62	0.46	0.59
Zn	0.093	0.090	0.083	0.12
Sr	0.080	0.074	0.062	0.089
Mo	0.033	0.029	0.036	0.054
Ba	0.56	0.30	0.30	0.38
Pb	0.16	0.13	0.15	0.065
H_2O	19.0	15.4	16.5	19.3
L.O.I.	—	—	—	—
Reference				
Reduced analyses				
Mn	33.8	31.6	29.4	38.3
Fe	13.6	10.1	17.4	15.0
Co	0.60	0.39	0.42	0.20
Ni	1.65	1.80	1.56	1.59
Cu	0.68	0.97	0.83	0.92
Pb	0.14	0.14	0.27	0.10

XXX (continued)

DWHD 47b	DWBG 78	Chal. 302	
S 41°59'	S 44°08'	S 42°43'	Station
W 102°01'	W 100°58'	W 82°11'	Latitude
4,200	4,100	2,156	Longitude
Calc. O.	Calc. O.	Calc. O.	Depth (m)
Dredge	Core	Trawl	Asso. sed.
—	1–2	—	Sampler
—	—	—	Nod. dia. (cm)
Out. layer	Whole nod.	—	Sp. G.
			Sect. sampled
		Chemical analyses (Weight percentages)	
4.90	3.2	2.8	Al_2O_3
10.06	8.8	11.8	SiO_2
—	—	—	P
0.86	—	0.2	K
2.39	1.5	1.7	Ca
0.39	0.35	0.4	Ti
24.2	19.5	12.9	Mn
9.9	9.9	19.4	Fe
0.20	0.10	0.08	Co
0.92	0.67	0.18	Ni
0.18	0.32	0.11	Cu
—	0.04	0.04	Zn
—	0.05	0.12	Sr
—	0.031	0.04	Mo
—	0.08	0.12	Ba
0.074	0.048	0.031	Pb
19.52	—	—	H_2O
—	33.0	27.0	L.O.I.
SKORNYA-	GOLDBERG	GOLDBERG	Reference
KOVA et al.	and MENARD	and MENARD	
(1962)	(pers. comm., 1960)	(pers. comm., 1960)	
			Reduced analyses
37.0	31.4	19.7	Mn
15.2	16.0	29.7	Fe
0.30	0.16	0.12	Co
1.41	1.08	0.28	Ni
0.27	0.52	0.17	Cu
0.11	0.077	0.047	Pb

TABLE

Station	Msn 85G	Msn 85G	V. 16–34
Latitude	S 57°43′	S 57°43′	S 54°30′
Longitude	E 169°12′	E 169°12′	W 163°19′
Depth (m)	5,288	5,288	4,540
Asso. sed.	R. C.	(Nodule	—
Sampler	Core	46 cm	Dredge
Nod. dia (cm)	2×2×3	below	1 cm Crust
Sp. G.	—	surface)	—
Sect. sampled	x-sect.	Whole nod.	x-sect.

Chemical analyses (Weight percentages)			
Al_2O_3	5.2	4.5	3.7
SiO_2	25.9	26.6	20.9
P	—	—	—
K	—	—	—
Ca	1.44	1.36	1.32
Ti	0.59	0.45	0.61
Mn	10.7	12.3	14.4
Fe	19.2	15.3	14.9
Co	0.17	0.15	0.30
Ni	0.19	0.23	0.37
Cu	0.10	0.17	0.15
Zn	0.047	0.044	0.057
Sr	0.11	0.10	0.10
Mo	0.011	0.007	0.017
Ba	0.44	0.38	0.42
Pb	0.15	0.15	0.17
H_2O	17.7	18.1	16.5
L.O.I.	—	—	—
Reference			

Reduced analyses			
Mn	20.8	24.2	20.9
Fe	37.4	30.1	21.6
Co	0.33	0.30	0.44
Ni	0.37	0.45	0.54
Cu	0.19	0.33	0.22
Pb	0.09	0.09	0.085

XXX (continued)

Msn 90G	Msn 91G	Msn 91G	Station
S 63°04′	S 64°11′	S 64°11′	Latitude
E 178°29′	W 165°56′	W 165°56′	Longitude
3,583	2,932	2,932	Depth (m)
Sil. O.	Sil. O.	(Nodule	Asso. sed.
Core	Core	11 cm	Sampler
1.5 mm Crust	0.2 cm Crust	below	Nod. dia. (cm)
—	—	surface)	Sp. G.
x-sect.	x-sect.	x-sect.	Sect. sampled
		Chemical analyses (Weight percentages)	
7.4	4.6	4.7	Al_2O_3
33.3	24.3	27.8	SiO_2
—	—	—	P
—	—	—	K
1.79	1.52	1.44	Ca
0.46	0.63	0.58	Ti
8.8	14.8	14.2	Mn
12.4	11.0	10.8	Fe
0.14	0.17	0.15	Co
0.27	0.70	0.68	Ni
0.13	0.29	0.36	Cu
0.058	0.10	0.12	Zn
0.11	0.10	0.10	Sr
0.005	0.011	0.012	Mo
0.25	0.39	0.36	Ba
0.14	0.15	0.15	Pb
12.2	17.4	—	H_2O
—	—	—	L.O.I.
			Reference
			Reduced analyses
18.7	27.6	28.2	Mn
26.3	20.5	23.0	Fe
0.30	0.32	0.30	Co
0.57	1.30	1.35	Ni
0.28	0.54	0.71	Cu
0.12	0.29	0.30	Pb

ganese nodules and the inclusion of all the elements found therein, will, indeed, be complicated. At least fifteen different chemical and physical agencies can be expected to have some effect on the formation and composition of this material. If biological effects are included, the number of agencies can be expected to be markedly increased. Among the more important of these agencies will be: pH and Eh of the ocean-floor environment, chemical and physical properties of the colloids, concentration of the colloids in sea water, concentration of the trace elements in sea water, concentration of detrital minerals in sea water, ion size, nature of the charge on the scavenger sols and the scavenged ions, charge densities of the ions, water currents, temperature and hydrostatic pressure, time and space relationships of the sources of the elements and the forming nodules, biological phenomenon, and rates of formation of the associated sediments.

Table XXX lists compositional data on sixteen elements in the nodules from 166 stations in the Pacific Ocean. Also listed in Table XXX are location, depth, associated sediments, sampling device, nodule size, specific gravity, and a description of the part of the nodule used for chemical analyses. The data of Table XXX are arranged in order of the location in the ocean from which the samples were obtained, starting with the northwesternmost corner of the Pacific Ocean and proceeding eastward in rows of 10° of latitude. The analyses of the nodules listed in Table XXX were generally performed on a cross-sectional or bulk sample of the nodule as a whole, including the nucleus.

To obtain a greater degree of uniformity in the analyses of the nodules, the assays for manganese, iron, cobalt, nickel, copper, and lead were reduced to a detrital mineral-free basis. Silica, alumina, water, and calcium carbonate, in amounts greater than 5%, were considered detrital minerals for purposes of computing the reduced analyses. The reduced analyses are also listed in Table XXX. Statistics on these reduced analyses are shown in Table XXXI.

Method of analysis

Analyses on nodules listed in Table XXX not referenced to

TABLE XXXI

METAL CONTENT AND RATIO STATISTICS OF THE VARIOUS COMPOSITIONAL REGIONS OF MANGANESE NODULES IN THE PACIFIC OCEAN

CHEMICAL COMPOSITION OF MANGANESE NODULES 223

Region	Statistic	Reduced weight percentages						Weight percentage ratios					
		Mn	Fe	Co	Ni	Cu	Pb	Mn/Fe	Mn/Ni	Mn/Pb	Ni/Cu	Fe/Co
A	Maximum	35.6	39.5	0.82	0.91	0.61	0.40	1.25	94.3	420	5.0	517
	Minimum	7.6	19.7	0.06	0.08	0.07	0.047	0.22	30.0	21	0.46	37
	Average	21.7	28.3	0.35	0.46	0.32	0.21	0.79	53.8	146	2.21	168
AD	Maximum[1]											
	Minimum	22.3	26.4	0.69	0.30	0.43	0.28	0.85	74.4	80	0.68	38
	Average											
B	Maximum	57.1	3.5	0.20	0.83	0.43	0.089	61.8	870	9,500	5.7	318
	Minimum	40.6	1.1	0.008	0.06	0.013	0.006	11.6	69	540	1.0	7
	Average	49.8	2.3	0.055	0.26	0.14	0.047	29.8	356	2,200	3.0	106
BC	Maximum	50.4	13.5	0.36	2.18	1.40	0.12	9.87	45.8	3,360	1.92	185
	Minimum	40.0	5.1	0.045	1.52	0.72	0.038	2.82	18.7	333	1.42	37
	Average	43.6	10.7	0.15	1.67	0.04	0.070	4.94	28.2	1,178	1.61	103
C	Maximum	45.3	25.8	0.75	2.37	2.92	0.52	6.21	59.4	608	2.58	260
	Minimum	15.0	7.3	0.076	0.54	0.44	0.049	0.97	11.3	51	0.64	19
	Average	33.3	17.7	0.39	1.52	1.13	0.18	2.14	23.7	261	1.51	60
CD	Maximum	38.3	24.4	1.22	1.94	1.46	0.54	3.41	45.5	860	1.74	38
	Minimum	23.4	10.1	0.47	0.59	0.34	0.04	1.09	15.9	43	0.97	8
	Average	31.7	17.5	0.69	1.45	1.09	0.29	1.96	25.4	201	1.36	27
D	Maximum	33.2	27.4	2.09	0.97	0.39	0.41	1.84	65.0	232	6.1	39
	Minimum	20.2	16.6	0.70	0.37	0.13	0.10	0.90	27.7	60	1.4	8
	Average	28.5	22.6	1.21	0.66	0.21	0.30	1.31	45.6	109	3.4	22
Statistics on all the samples													
	Maximum	57.1	39.5	2.09	2.37	2.92	0.54	61.8	870	9,500	6.1	517
	Minimum	7.6	1.1	0.008	0.06	0.013	0.006	0.22	11.3	21	0.46	8
	Average	32.4	18.5	0.47	1.14	0.80	0.19	4.11	60.0	431	1.95	74

[1] Only one sample point in region AD-1.

another source were performed by the author. The method used was an X-ray fluorescence technique developed by the author and Mr. George M. Gordon of the Department of Mineral Technology of the University of California.

The equipment used was a General Electric X R.D.–5 with a tungsten tube. Lithium fluoride was used as the diffraction element in assaying for all elements above calcium in the atomic table and E.D.D.T. was used in conjunction with a helium path for all elements with an atomic number less than calcium. Flow counters were used in conjunction with a pulse-height analyzer to eliminate X-ray lines of different, but integral, orders in gathering count data. The stability of the equipment was found to be excellent.

The equipment was calibrated by the use of standard ores made from pure-oxide forms of the elements in the nodules and carefully mixed in proportion to the amounts of these elements generally found in the manganese nodules. Chemically analyzed standards of the nodules themselves were also used. As a final check, a known amount of the element in question was added to selected samples of the nodules and careful counts were taken on these samples before and after the addition of the extra amount of the element.

The method involved the determination and subsequent use of absorption and activation factors for the lines of the various elements. All the absorption and activation factors were carefully determined using the standard ores. When applied to chemically analyzed samples of the nodules these methods were accurate to three significant figures in most cases which was within the limits of the accuracy of the chemical analyses performed on our standards.

Because of the generally small grain size of the crystallites in the nodules and the generally uniform distribution of the elements in the nodules, they yield very well to analysis by X-ray fluorescence techniques. Adequate sample preparation, which is an almost impossible task when dealing with large-grained terrestial ores, is easily attained with the manganese nodules. Different cuts of the same sample always yielded closely reproducible assays by the method used.

Regional variations in the composition of the manganese nodules

If the assay data of Table XXX are plotted on a map of the Pacific Ocean, definite regional variations in the composition of the nodules are noticeable. Fig. 63 shows the approximate boundaries of the compositional regions noted and also shows the locations from which samples of the nodules were obtained and analyzed to provide the data of Table XXX.

Although the 166 samples listed in Table XXX are hardly adequate sampling when dealing with the area of the Pacific Ocean, these data, considering the uniform nature of pelagic sediments over large areas, are significant. The boundaries of the proposed compositional regions of the nodules in the Pacific are, of course, somewhat tentative. Anomalies probably exist within each region. When these regions were first noticed (MERO, 1960b), the high-manganese area was shown as a continuous belt paralleling the North and South American coasts. The more recent data included in this volume has reduced this belt to several separate regions. Additional data can be expected to modify the boundaries shown in Fig. 63.

The basis for determining whether a nodule is "high" in regard to the percentage of a particular element or not is, of course, rather arbitrary. In this analysis, however, a weight percentage of about 1% on a detrital mineral-free basis was chosen to indicate "high" metal contents in the case of cobalt, nickel, and copper, while a weight percentage above 40% was considered "high" in the case of manganese. If the manganese/iron ratio was less than unity, the nodule was considered "high" in iron.

A-regions (high iron)

The nodules in the areas labeled, A, in Fig. 63 are characterized by a manganese/iron ratio generally less than one. The iron/cobalt ratios are higher than those of the other regions, on the average, and range as high as 517. The A regions generally lie along the continents, however, one, A-3, lies in the area between New Zealand and Tahiti. The average assays of the nodules from the A regions, on a detrital mineral-free basis, are: iron 28.3%, manganese 21.7%, cobalt 0.35%, nickel 0.46%, copper 0.32%, and lead 0.21%.

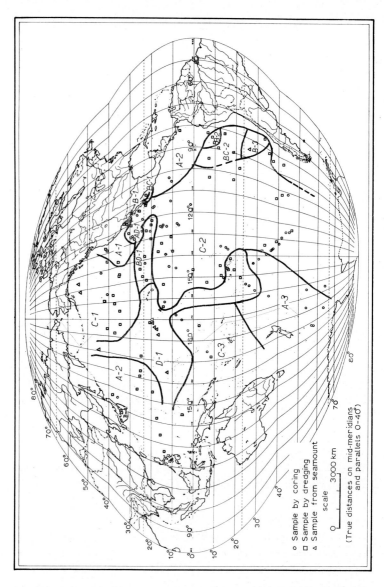

Fig. 63. Map of the Pacific Ocean, showing the compositional regions of the manganese nodules in this ocean. Nodules from the *A* regions are high in iron; from the *B* regions, high in manganese; from the *C* regions, high in nickel and copper; from the *D* regions, high in cobalt. Also indicated on this map are the locations of points where samples of nodules have been obtained on which chemical analyses have been performed. The assay data are listed in Table XXX.

KRAUSKOPF (1957) has outlined a process by which iron and manganese can be differentially precipitated from sea water. Generally, the iron is dropped first. With such a process operating along the coasts of South and North America on the iron and manganese coming off the continents, iron would drop first in the near-shore A regions leaving the waters comparatively rich in manganese for the B and C regions. Presumably, then, the same process would account for the high-iron zones all around the Pacific basin. The high-iron region in the area between New Zealand and Tahiti, thus, might be a reflection of a current which would be carrying the iron toward Tahiti.

B-regions (high manganese)

Near the west coasts of North and South America are three areas where the nodules are characterized by a very high manganese/iron ratio. This ratio ranges between 12–60 and averages 30 in these areas, whereas the average for the whole of the Pacific Ocean for this ratio is 4.1. The iron/cobalt ratio is generally high, averaging 116 for six samples within these three areas. The most northerly "high" manganese area lies in and near the Gulf of California. The southern two regions are centered on latitudes, 10° south and 30° south. There apparently is a transition zone between the two southern regions and the C region. In this transition zone the nodules show compositional characteristics of both zones. One sample, Alb. 4660 from the central region, assayed almost 80% MnO_2 on a dry-weight basis. Another interesting sample, Alb. 4711, in the transition zone, BC-2, assayed 43% manganese, 1.8% nickel, and 0.7% copper on a dry-weight basis. The average assays, on a detrital mineral-free basis, for the B regions are: manganese 49.8%, iron 2.3%, cobalt 0.055%, nickel 0.26%, copper 0.14% and lead 0.047%.

To explain the formation of these "high" manganese zones of the nodules, the process outlined by KRAUSKOPF (1957), mentioned earlier in this section, might be used. Thus, iron being precipitated in the A regions would leave the waters impoverished in this elements seaward of these zones. Thus we might expect to find manganese-rich zones bordering the A zones at all locations. The data, however, do not indicate such to be the case. KRAUSKOPF (1957) also discusses processes which involve the differential precipitation of manganese and iron leached from volcanic rocks. This process, or an eruptive deficient in iron relative to manganese, is most likely the cause of the mangnese-rich zones in the eastern South Pacific.

Generally, the nodules in the high-manganese regions show a very low content of cobalt, nickel, and copper. A very rapid rate of formation seems to be the best explanation for this situation. If the manganese colloids do not have sufficient time in contact with sea water between the time they precipitate from the water and that when they are agglomerated into the nodules they would not be able to scavenge much of a load of the guest metals.

Manganese-rich nodules found off the southeast coast of Japan (NIINO, 1959) are apparently the result of a rapid precipitation and agglomeration of manganese from manganese-rich springs entering the ocean floor near these deposits. These nodules and crusts lie in the Fuji volcanic zone and the eruptives exposed in these areas could also be a source of the manganese in these nodules. There is a relatively high percentage of calcium carbonate in the nodules from this area as is common in nodules formed in relatively shallow waters. Other than an unusually low silica and alumina content, the nodules in this area off Japan are quite similar in composition to the nodules from the other B regions.

Bacterial action has been shown to precipitate preferentially manganese and iron from solution in sea water (LJUNGGREN, 1953), and this process might be an explanation for the anomalous composition of the nodules in the B regions.

C-regions (high nickel and copper)

The parts of the Pacific Ocean, farthest removed from land, both continental and island, seem to be regions of relatively high nickel–copper nodules. These regions, labeled C in Fig. 63, are, by far, the dominant compositional regions in the Pacific Ocean. The manganese/iron ratio of the C regions is relatively stable, ranging between 1–6 and averaging 2.1. The nickel/copper ratio of the nodules from these regions have a much smaller range than in the other regions. The iron/cobalt ratio exhibits a wide range of values, between 19–260 and averaging 60. The averages of the essays, on a detrital mineral-free basis, of the nodules from the C regions are: manganese 33.3%, iron 17.7%, cobalt 0.39%, nickel 1.52%, copper 1.13%, and lead 0.18%.

The copper assays of the nodules from this region show a greater range in value than do the nickel assays. REVELLE et al. (1955) have shown a biological cause for part of the copper in pelagic sediments and such processes may also be responsible for placing part of the copper in the manganese nodules. On the average, the copper content tends to increase near the equator if the nodules of the "high" cobalt regions are not considered. The high content of other elements such as barium, which is also precipitated through the action of biotic agencies, in the sediments under the productivity zone of the equatorial divergence has been noted by several authors (GOLDBERG and ARRHENIUS, 1958).

Of all the metal content ratios calculated from the data of Table XXX, the manganese/nickel ratio seems to be the most consistent, if the anomalous values of the B regions are disregarded. The lead content in certain of the nodules of the C regions is low relative to the overall value of the manganese/nickel ratio. Various rates of growth of the nodules in the C regions from location to location are thus inferred.

D-region (high cobalt)

Centered on topographic highs, in the central part of the Pacific Ocean, is a region in which the manganese nodules assay relatively high in cobalt. Nodules from this region average about 1.2% cobalt. The range of cobalt values is between 0.7–2.1%. The averages of assays, on a detrital-mineral-free basis, for other elements in the nodules of this region are: manganese 28.5%, iron 22.6%, nickel 0.66%, copper 0.21%, and lead 0.30%. The manganese/iron ratios of these regions varied less than in any of the other regions, ranging between 0.9–1.8 and averaging 1.3. The manganese/nickel ratio is also relatively stable, showing a smaller range of values than the other regions. The iron/cobalt ratio also is more stable in these regions, ranging between 8–39 and averaging 22.

The relationship of high-cobalt nodules and topographic highs in mid-ocean is marked. Nodules relatively high in cobalt, however, are found in other environments of the oceans, both in the basins and on the continental rises.

The high percentage of lead in the nodules and crusts from region D is notable as is the low copper content. Possibly, the rocks forming the seamounts in these regions are relatively rich in cobalt, thus, the water around the seamounts would also be rich in cobalt. Or possibly it may be that a highly oxidizing environment promotes the incorporation of the cobalt ion in the precipitating sols or in the nodules. The areas of topographic rises in the ocean are generally characterized by highly oxidizing environments because of the high water current velocities over these features.

Probably the most interesting area noted on the map shown in Fig. 63 is the transition zone CD-1 which extends eastward along the north 22° line of latitude. It is an area in which a great many of the nodules seem to have a relatively high-cobalt composition, but also contain appreciable amounts of copper and nickel. Part of this zone seems to lap over into region A-1 forming an AD zone. This latter zone is, however, predicated on only one sample, DWBD-1, which may be anomalous.

Other transition zones

Although not shown in Fig. 63, other transition zones between the boundaries of the A, B, C, and D regions can be expected to be found. Only where analyses on actual samples showed the zones to exist, were they included in Fig. 63. Parts of regions A-3 and A-4 could be classed as C–D transition zones. The manganese/iron ratios less than 1 and the nickel/cobalt ratios greater than 1 of the samples in these areas, however, caused them to be classified as A regions.

Analyses of various shells within a single nodule

Table XXXII shows the analysis of various shells within a manganese nodule from a station at N 9°57′, W 137°47′, in the C-2 region of the eastern Pacific. Although small changes in the concentration of certain elements, such as in the amount of cobalt increasing in value from the core of the nodule toward the outside shell and copper decreasing in value toward the core, can be noticed, the composition of the nodules, from this station at least, seems to be generally uniform throughout the various shells of a single nodule.

Closely spaced sampling of a manganese nodule deposit

As far as is known, only one area of the ocean has been sampled to any great degree for manganese nodules and from which the data are available. This area, centered about 370 km southwest of the southern tip of Baja California, is shown in Fig. 64. Ten samples were taken this area along a track which started at N 22°00′, W 116°14′, about 560 km west of Cape San Lucas, Baja California, and ended at N 21°53′, W 112°47′, about 240 km west of the Cape. Nodules were obtained from every station at which dredging was done along this track.

The appearance of the nodules from these stations did not vary markedly. The size of the nodules from these ten stations ranged

TABLE XXXII

CHEMICAL COMPOSITION OF SHELLS WITHIN A MANGANESE NODULE[1]

Section sampled	Weight percentages			
	1–2 cm from core	2–3 cm from core	3–4 cm from core	4–5 cm from core
SiO_2	14.7	14.7	12.3	12.5
Ca	1.23	1.46	1.56	1.40
Ti	0.33	0.52	0.61	0.38
Mn	30.7	29.1	27.7	32.1
Fe	2.3	3.9	5.3	3.2
Co	0.11	0.16	0.22	0.20
Ni	1.37	1.26	1.33	1.05
Cu	1.34	1.00	1.00	0.77
Sr	0.051	0.051	0.056	0.051
Ba	0.56	0.46	0.45	0.33
H_2O	12.7	15.9	14.9	11.5

[1] Manganese nodule from station Alb. 13 at N 9°57′, W 137°47′, depth 4,930 m. This nodule was about 10 cm in diameter. The core was not sampled.

between about 2–10 cm in diameter and averaged 5 cm. The nodules generally were shaped like closed clamshells at all stations except station DH-2 where they were roughly spherical and station DH-10 where a few of the nodules were cylindrical. All these nodules were resting on a red clay sediment.

Table XXXIII lists the coordinates of the stations from which these samples were recovered and gives chemical analyses for the manganese, iron, cobalt, nickel, and copper of the nodules. There is a marked decrease in the percentage of iron in the nodules and slight decreases in the percentages of cobalt, nickel, and copper as the mouth of the Gulf of California is approached. Conversely, the

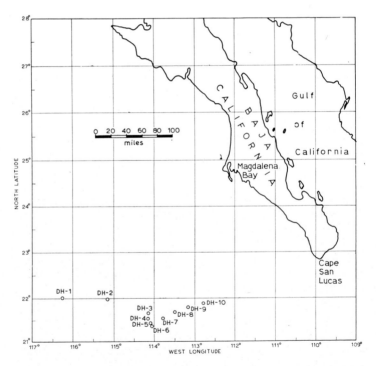

Fig. 64. Map of the area of the Pacific Ocean, west of the southern tip of Baja California, showing the locations where manganese nodules were dredged from a series of relatively closely-spaced stations.

CHEMICAL COMPOSITION OF MANGANESE NODULES 233

TABLE XXXIII

CHEMICAL ANALYSES OF MANGANESE NODULES FROM A SERIES OF CLOSELY SPACED STATIONS

Station	DH 1	DH 2	DH 3	DH 4	DH 5	DH 6	DH 7	DH 8	DH 9	DH 10
Latitude	N 22°00′	N 21°50′	N 21°40′	N 21°31′	N 21°27′	N 21°21′	N 21°33′	N 21°40′	N 21°48′	N 21°53′
Longitude	W 116°14′	W 115°12′	W 114°11′	W 114°08′	W 114°07′	W 114°06′	W 113°48′	W 113°30′	W 113°03′	W 112°47′
Depth (m)	3,480	3,430	3,800	3,800	3,800	3,660	3,660	3,420	3,450	3,385
Asso. sed.	R. C.	R. C.	R. C.	R. C.	R. C.	R. C.	R. C.	R. C.	R. C.	R. C.
Sampler	Dredge	Dredge	Dredge	Dredge	Dredge	Dredge	Dredge	Dredge	Dredge	Dredge
Nod. dia. (cm)	2–8	1–8	2.5–7.5	2–9.5	1.5–13.5	1.5–10	2–8	3–10	1.5–7.5	2–6
Sect. sampled	x-sect.	1 cm nod.	x-sect.	x-sect.	Whole nod.	0.5 nod.	Whole nod.	x-sect.	0.5 nod.	Whole nod.
Chemical analyses (Weight percentages)										
Mn	27.8	25.1	28.8	28.2	27.6	28.9	24.4	28.9	31.0	30.3
Fe	10.4	13.0	9.9	8.1	10.9	9.0	7.5	9.4	8.3	5.7
Co	0.08	0.11	0.09	0.05	0.11	0.08	0.05	0.07	0.04	0.01
Ni	1.02	1.02	1.19	1.46	1.23	1.35	1.24	1.24	1.10	0.54
Cu	0.61	0.50	0.64	0.77	0.62	0.72	0.62	0.60	0.47	0.31
H_2O	31.2	33.4	33.1	34.5	33.5	34.3	30.9	34.2	32.9	36.1

nodules show a slight increase in the amount of manganese as the Gulf is approached. These trends in the composition of the nodules were expected as the Gulf of California, on the basis of several samples of nodules taken within this body of water, had been classed as "high" manganese.

The water content of these nodules is somewhat higher than the average of that listed in Table XXX. These samples were stored in water-proof plastic bags until the time when they were analyzed. Some of the nodules on which analyses were run which are listed in Table XXX were stored in the attics of museums for the past 60 years. Much of the water of hydration must have evaporated from those samples during that time.

AMOUNTS OF VARIOUS METALS IN THE MANGANESE NODULES

Based on the tonnage calculations of Table XXV and the average compositions as shown in Table XXVIII, calculations can be made of the amounts of various elements in the manganese nodules at the surface of the Pacific Ocean sediments. These amounts are listed in Table XXXIV. Even if only about 1% of the nodules in the Pacific Ocean prove economic to mine, the reserves of many metals in the nodules will still be measured in terms of thousands of years at the present rates of free world consumption.

Also listed in Table XXXIV are the ratios for the concentration of various elements in the nodules to that of the same elements in sea water. In the case of manganese, this element is 240 million times more concentrated in the manganese nodules than it is in sea water.

MANGANESE NODULES IN THE ATLANTIC AND INDIAN OCEANS

Manganese nodules have been found at a number of locations in the Atlantic and Indian Oceans. Deposits of the nodules in these oceans, however, do not seem to be as widespread as in the Pacific Ocean. Fewer deposits of the nodules in the Atlantic Ocean might

TABLE XXXIV

AMOUNT AND RATE OF ACCUMULATION OF VARIOUS ELEMENTS IN THE PACIFIC OCEAN MANGANESE NODULES

Element	Average percentage in nodules (%)[1]	Amount of element in nodules of Pacific Ocean (10^9 tons)[2]	Annual rate of accumulation of element into the nodules (10^6 tons)[3]	Concentration of element sea water, (p.p.b.)[4]	Conc. ratio (Conc. in nodules/conc. in sea water)
B	0.029	0.48	0.0018	4,500	64
Na	2.6	43.0	0.16	$1.1 \cdot 10^7$	2.5
Mg	1.7	28.2	0.11	$1.3 \cdot 10^6$	13
Al	2.9	48.0	0.18	10	$2.9 \cdot 10^6$
Si	9.4	156.0	0.56	—	—
K	0.8	13.2	0.048	$3.8 \cdot 10^5$	25
Ca	1.9	31.2	0.12	$4 \cdot 10^5$	—
Sc	0.001	0.017	0.0006	0.04	$2.5 \cdot 10^5$
Ti	0.67	11.1	0.041	1	$7 \cdot 10^6$
V	0.054	0.90	0.0033	3	$2 \cdot 10^5$
Cr	0.001	0.017	0.0006	0.05	$2 \cdot 10^5$
Mn	24.2	400	1.5	1[5]	$2.4 \cdot 10^8$
Fe	14.0	232	0.84	1[6]	$1.4 \cdot 10^8$
Co	0.35	5.8	0.022	0.4	$9 \cdot 10^6$
Ni	0.99	16.4	0.061	2	$5 \cdot 10^6$
Cu	0.53	8.8	0.033	2[5]	$3 \cdot 10^6$
Zn	0.047	0.78	0.0029	5	10^5
Ga	0.001	0.017	0.0006	0.03	$3 \cdot 10^5$
Sr	0.081	0.13	0.005	8,000	10^2
Y	0.033	0.55	0.001	0.3	$5 \cdot 10^5$
Zr	0.063	1.04	0.004	—	—
Mo	0.052	0.86	0.003	12	$4 \cdot 10^4$
Ag	0.0003	0.005	0.00002	0.15	$2 \cdot 10^4$
Ba	0.18	3.0	0.011	50	$3 \cdot 10^4$
La	0.016	0.26	0.001	0.3	$5 \cdot 10^5$
Yb	0.0031	0.051	0.0002	—	—
Pb	0.09	1.5	0.0034	5	$2 \cdot 10^5$

[1] Dry-weight percentages as listed in Table XI.
[2] Amounts in metric tons based on a total tonnage of $1.66 \cdot 10^{12}$ as indicated in Table VI.
[3] Amounts in metric tons based on an annual rate of accumulation of the nodules $6.0 \cdot 10^6$ metric tons.
[4] p.p.b. = parts per billion. Data as compiled by RILEY and SINHASENI (1958).
[5] Variable.
[6] Highly variable.

be expected in view of the much higher rates of sedimentation in this ocean. Fig. 65 shows the locations where nodule deposits have been sampled in the Atlantic. Also shown in Fig. 65 are the concentration estimates from photographs, showing nodules. Table XXXV lists a number of analyses on nodules from the Atlantic.

No regional compositional trends of the nodules are evident in the Atlantic Ocean from the data of Table XXXV. The composition of the nodules from this ocean generally conforms to that of the A regions in the Pacific. Only one sample of nodules was available from the South Atlantic, the other four samples on which analyses were run from this part of the Atlantic were either crusts or manganese-dioxide impregnated pumice.

An interesting deposit of manganese nodules in the Atlantic is that on the Blake Plateau. The Blake Plateau is an area of about 200,000 km^2 off the coasts of the Carolinas and Florida and it ranges in depth between 200–1,000 m. The photograph of nodules on the Blake Plateau, shown in Fig. 60, indicates a relatively heavy concentration of nodules in this area. The apparent reason manganese nodules are forming on the Blake Plateau, in relatively shallow water and close to a continental shore, is that the Gulf Stream flows through this area and the high water currents apparently sweep it free of terrigenous sediments. The calcium-phosphate content of the nodules from the Blake Plateau is relatively high as compared with nodules from other locations in either the Atlantic or Pacific Oceans. Samples W.H.O.I., A266-40, A266-45, and A266-41, as listed in Table XXXV, are from the Blake Plateau.

Manganese nodules in the Indian Ocean

Samples of nodules from the Indian Ocean show about the same average composition as the nodules from the Atlantic Ocean. Nodules from an area south of Australia, however, assay relatively high in copper, to 1.8% (GOLDBERG, 1954). Table XXXVI shows the assays for six samples of nodules from the Indian Ocean.

NODULES IN ATLANTIC AND INDIAN OCEANS 237

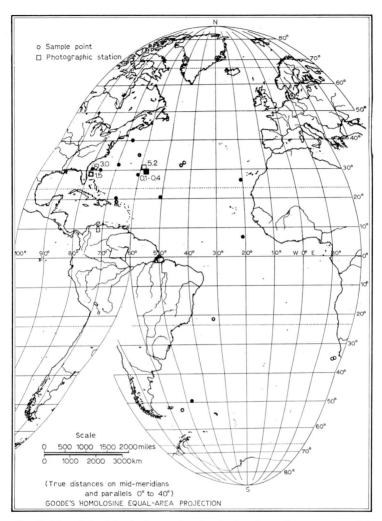

Fig. 65. Map of the Atlantic Ocean, showing the locations of manganese-nodule sample points and of photographic stations. The numbers by the photographic stations indicate the concentration of nodules, in g/cm², on the sea floor at that point.

TABLE XXXV

CHEMICAL ANALYSES OF MANGANESE NODULES FROM THE ATLANTIC OCEAN

Station	Blake T	U.S. Navy	WHOI	A.266-40	A.266-45	A.266-41	Theta 1-7
Latitude	N 39°57'	N 30°	N 30°51'	N 30°53'	N 30°57'	N 30°59'	N 32°13'
Longitude	W 66°49'	W 76°	W 78°27'	W 78°47'	W 78°21'	W 78°14'	W 69°06'
Depth (m)	3,710	2,645	732	815	810	879	5,290
Asso. sed.	T. Sed.	Calc. O.	T. Sed.	Coral S.	Coral S.	Coral S.	R. C.
Sampler	Trawl	Dredge	Dredge	Dredge	Dredge	Dredge	Dredge
Nod. dia. (cm)	1–4	1–2	4–8	$2 \times 10 \times 13$	$2.5 \times 8.5 \times 18$	$8 \times 6 \times 6$	$0.5 \times 0.5 \times 3$
Sp. G.	2.42	2.48	2.67	—	—	—	3.19
Sect. sampled	Whole nod.	Whole nod.	x-sect.	x-sect.	x-sect.	x-sect.	x-sect.
Chemical analyses (Weight percentages)							
Al_2O_3	6.9	5.3	4.9	4.8	2.7	4.5	9.8
SiO_2	27.1	7.5	2.9	2.9	2.3	4.9	25.2
P	0.028	0.051	0.147	—	—	—	0.169
K	0.56	0.30	0.31	—	—	—	0.95
Ca	0.89	1.94	7.32	9.9	11.1	8.9	1.17
Ti	0.57	0.68	0.34	0.20	0.19	0.21	0.37
Mn	10.1	14.8	15.7	11.8	11.1	13.6	14.8
Fe	16.9	20.0	15.5	14.3	14.1	10.4	9.3
Co	0.12	0.40	0.41	0.39	0.38	0.48	0.14
Ni	0.21	0.27	0.59	0.42	0.33	0.53	0.29
Cu	0.10	0.10	0.14	0.06	0.03	0.10	0.30
Zn	0.069	0.067	0.055	0.049	0.039	0.058	0.043
Sr	0.086	0.13	0.19	0.13	0.18	0.17	0.058
Mo	0.035	0.43	0.055	0.035	0.036	0.040	0.024
Ba	0.50	0.62	0.52	0.41	0.34	0.46	0.37
Pb	0.11	0.12	0.15	0.21	0.19	0.15	0.14
H_2O	17.3	23.3	15.6	12.5	12.3	11.0	13.5
L.O.I.	—	—	—	—	—	—	—
Reference							

TABLE XXXV (continued)

Station	Muir Smt	SW 270	Theta 1-6	BM 198a	WHOI 7	BM 197	V.15-151
Latitude	N 34°52'	N 28°05'	N 29°17'	N 29°18'	N 30°49'	N 31°49'	N 20°24'
Longitude	W 62°30'	W 60°49'	W 57°23'	W 57°20'	W 44°33'	W 43°25'	W 66°24'
Depth (m)	1,460	5,760	5,840	5,781	3,540	3,700	5,520
Asso. sed.	Coral D.	R. C.	R. C.	—	Calc. O.	—	R. C.
Sampler	Dredge	Core	Dredge	—	Dredge	—	Dredge
Nod. dia.(cm)	2 cm Crust	2–3	8×9×10	—	0.1	—	4–5
Sp. G.	—	—	2.40	—	—	—	2.33
Sect. sampled	x-sect.	x-sect.	Out. 1 cm	—	Whole nod.	—	x-sect.
Chemical analyses (Weight percentages)							
Al$_2$O$_3$	2.6	7.1	8.2	—	5.0	—	6.6
SiO$_2$	2.0	12.3	13.5	17.6	5.5	9.2	14.1
P	—	—	0.096	—	—	—	—
K	—	0.49	0.51	—	—	—	0.42
Ca	2.23	1.27	0.41	—	2.03	—	1.07
Ti	0.57	0.60	0.65	0.48	0.87	0.91	0.62
Mn	16.9	18.9	16.2	—	11.5	—	12.9
Fe	18.9	1.54	15.9	17.5	25.9	24.3	19.8
Co	0.91	0.33	0.54	0.30	0.79	0.48	0.24
Ni	0.27	0.57	0.49	0.36	0.16	0.19	0.27
Cu	0.04	0.26	0.28	0.20	0.07	0.10	0.19
Zn	0.053	0.071	0.067	—	0.051	—	0.043
Sr	0.13	0.080	0.077	—	0.09	—	0.088
Mo	0.039	0.038	0.031	0.047	0.017	0.032	0.030
Ba	0.71	0.80	0.69	—	0.67	—	0.69
Pb	0.23	0.10	0.13	—	0.20	—	0.16
H$_2$O	21.5	21.3	18.1	—	17.1	—	19.6
L.O.I.	—	—	—	12.5	—	12.9	—
Reference				Willis and Ahrens (1962)		Willis and Ahrens (1962)	

240 THE DEEP-SEA FLOOR

TABLE XXXV (continued)

Station	Chal. 16	UNK BH1	V.15-135	A. 322	A. 316	V.15-125	V.18-11
Latitude	N 20°39'	N 6°08'	S 20°59'	S 34°36'	S 34°42'	S 49°21'	S 53°00'
Longitude	W 50°33'	W 21°07'	W 31°49'	E 17°00'	E 16°54'	W 47°45'	W 52°54'
Depth (m)	4,450	7,180	4,180	2,743	3,200	4,980	3,060
Asso. sed.	Calc. O.	Calc. O.	—	—	—	—	Volc. R.
Sampler	Dredge	Tele. cable	Trawl	—	—	Trawl	Trawl
Nod. dia. (cm)	2–3	20×30	—	—	1 cm Crust	3×3×3	1 cm Crust
Sp. G.	—	2.46	—	—	—	—	—
Sect. sampled	Whole nod.	Out. 2 cm	Whole nod.	—	x-sect.	0.5 nod.	x-sect.

Chemical analyses (Weight percentages)

Al_2O_3	4.0	7.6	10.4	—	—	7.6	5.1
SiO_2	8.8	28.4	54.9	11.4	16.9	33.8	29.5
P	—	—	—	—	—	—	—
K	—	0.41	—	—	—	—	—
Ca	1.24	0.98	1.46	—	—	1.22	1.19
Ti	—	0.42	0.31	0.053	0.093	0.29	0.25
Mn	18.5	10.7	2.4	—	—	12.7	10.8
Fe	24.3	16.1	13.0	9.3	15.5	12.0	15.8
Co	—	0.30	0.09	0.20	0.20	0.09	0.15
Ni	—	0.24	0.013	0.71	0.75	0.46	0.39
Cu	—	0.14	0.04	0.083	0.095	0.09	0.08
Zn	—	0.052	0.035	—	—	0.063	0.067
Sr	—	0.082	0.11	—	—	0.084	0.080
Mo	—	0.030	0.000	0.033	0.044	0.015	0.014
Ba	—	0.50	0.14	—	—	0.25	0.30
Pb	—	0.093	0.16	—	—	0.13	0.12
H_2O	13.6	20.1	5.3	—	—	13.7	13.0
L.O.I.	—	—	—	13.6	13.4	—	—
Reference	Murray and Renard (1891)			Willis and Ahrens (1962)	Willis and Ahrens (1962)		

TABLE XXXVI

CHEMICAL ANALYSES OF MANGANESE NODULES FROM THE INDIAN OCEAN

Station	BM 966	V.16-17	V.16-19	V.16-20	V.16-29	Chal. 160
Latitude	N 6°55'	S 26°54'	S 29°52'	S 30°38'	S 37°50'	S 42°42'
Longitude	E 67°11'	E 56°04'	E 62°36'	E 70°07'	E 124°30'	E 134°10'
Depth. (m)	4,793	4,855	4,396	3,958	5,518	4,760
Asso. sed.	—	—	—	—	—	R. C.
Sampler	—	Trawl	Trawl	Trawl	Trawl	Trawl
Nod. dia.(cm)	—	1-5	3-9	1-2	1-6	—
Sp. G.	—	—	—	—	—	—
Sect. sampled	—	Whole nod.	x-sect.	x-sect.	0.5 nod.	—
Chemical analyses (Weight percentages)						
Al_2O_3	—	10.6	6.1	5.9	7.5	4.0
SiO_2	20.3	37.1	20.3	18.4	25.3	25.0
P	—	—	—	—	—	0.26
K	—	—	—	—	—	—
Ca	—	1.80	1.43	1.47	1.04	—
Ti	0.36	0.40	0.65	0.56	0.25	0.31
Mn	—	11.1	9.5	11.9	18.4	21.0
Fe	14.3	9.7	20.9	19.2	10.3	9.9
Co	0.15	0.21	0.29	0.25	0.15	0.09
Ni	0.95	0.42	0.13	0.24	0.90	1.15
Cu	0.40	0.29	0.11	0.16	0.45	1.81
Zn	—	0.055	0.047	0.060	0.065	—
Sr	—	0.093	0.096	0.093	0.064	—
Mo	0.054	0.008	0.013	0.013	0.023	—
Ba	—	0.26	0.50	0.45	0.27	—
Pb	—	0.13	0.15	0.15	0.10	—
H_2O	—	11.2	15.6	19.6	14.6	20.4
L.O.I.	14.2					
Reference	Willis and Ahrens (1962)					Goldberg (1954), Murray and Renard (1891)

CHAPTER VII

OCEAN MINING METHODS

Nature is a formidable adversary for the mining engineer on land; at sea, we are faced with nature at her most capricious level. The ever changing moods of the sea present a continual challenge to the ocean miner who wants her secrets and her riches. Waves, unchecked winds, salt-water corrosion, and a constantly shifting foundation make the design of an effective ocean mining method truly challenging.

While a myriad of methods exist for the mining of continental deposits, we find in the sea a greater uniformity in the character of the mineral deposits and their environments; thus, relatively few mining methods are needed. The methods described in this chapter can be used to exploit a relatively large number of individual deposits. A complete description of ocean-mining methods would require a volume in itself and only the general aspects of the technical and economic factors are herein described. Such discussion is germane at this point if we are to classify the sea-floor deposits as either mineral resources or potential mineral resources.

While some of the methods described, such as shallow water ladder-bucket dredges, have proved economic through decades of successful application, other methods have yet to be tested in a commercial operation, especially those proposed for deep-sea mining (MERO, 1959, 1960a). Hundreds of thousands of dollars have been spent in feasibility studies and in design studies of deep-sea mining methods by many independent agencies including industrial corporations and all have yielded favorable results. Because of the favorable conclusions of these detailed technical and economic studies, it is permissible, I believe, to term the phosphorite and manganese nodules, mineral resources. And if these studies indicate the technical feasibility of mining the manganese nodules, it is at least, technically, feasible to

THE MINING OF MARINE BEACHES AND OFFSHORE PLACERS

Emerged beach deposits can be mined with open-pit methods. Draglines are commonly employed as they can be used to work in the surf zone also. Almost any practical method of picking materials up off the ground is applicable to above sea-level beach deposits.

Offshore beach and placer deposits are mined by basically three methods, wire-line, bucket-ladder, and hydraulic dredges. Included in the wire-line methods are drag-bucket and clamshell-dredges and in hydraulic methods, suction and air-lift dredges. Fig. 66 illustrates the various methods.

Each method has its limitations and advantages. The hydraulic dredge has a greater production capacity in relation to the capital invested, especially if the dredged materials are to be moved any distance from the dredging site. The bucket-ladder or bucket-line type of dredge has a greater digging ability, and, in placers where the most valuable part of the deposit lies at the contact between the overburden and bedrock, this method is generally the most practical to

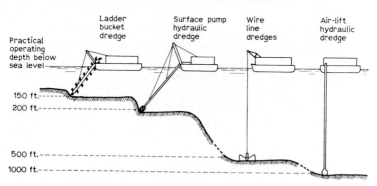

Fig. 66. Schematic diagram, showing the various dredging methods used in the mining of offshore placer deposits. The dredging depths indicated are the maximum practical economic working depths possible with the various methods.

use. Also, the ladder-bucket dredge requires less horsepower per unit of solids mined (ROMANOWITZ, 1962).

The ladder-bucket dredge is severely limited to the depth below water level at which it can work; so also is the hydraulic dredge when

Fig. 67. Deep-sea drag dredge used to recover tonnage samples of manganese nodules from the ocean floor. The bucket shown is 6 × 2 × 8 ft. There is about a ton of manganese nodules in the dredge bucket as shown.

the deposit is semi-compacted and must be broken up with cutter heads before entering the suction line. Both methods are impractical to use in offshore areas where wave motions and water currents can cause damage to the digging mechanism at the sea floor. Wire-line methods, on the other hand, are well suited for working in areas of high-current velocities or areas with high wave and swell motions.

Wire-line methods

In the wire-line methods, digging tools or buckets are suspended on a steel cable and lowered to the sediment surface where they are loaded and retrieved. Because of its simplicity, the drag-bucket is generally used in sampling sea-floor rock deposits in all depths up to and exceeding 30,000 ft. A deep-sea drag bucket is shown in Fig. 67. Because of the lack of control of placement of the bucket on the bottom and the necessity of dragging it over the sediment to fill it, which motion is inconvenient when operating from a dredge barge in depths exceeding about 50 ft., this method has not been used for offshore-mining applications. It is frequently used in underwater-dredging applications on the continents where channels are being dredged free of silt or other material. This method could be used in mining surficial deposits such as phosphorite nodules in relatively shallow water (less than about 1,000 ft.), however, to gain even minimal efficiencies in cleaning the deposit, a guide-wire system must be used to control the movement of the bucket from the surface to the ocean floor. Guide-wire methods have been proposed and considered practical by offshore drilling operators who presently anchor and move drilling barges with a system involving several cables laid off each end of the drilling vessels.

Grab buckets, including clamshells, orange peels, and other variations, consist of a digging device which in closing bites into the sediment and contains it inside the closed shell. The bucket and load are then hoisted to the surface where the shell is opened to dump the load. This method of mining is illustrated in Fig. 68. Clamshell buckets have been in use for about a hundred years. Today, probably more material, from a tonnage standpoint, is moved by this method than by any other single digging tool. Production rates of 800–1,000

Fig. 68. A clamshell-type of wire-line mining dredge. This dredge is used by the Yawata Iron and Steel Company to mine magnetite from the floor of Ariake Bay off Japan.

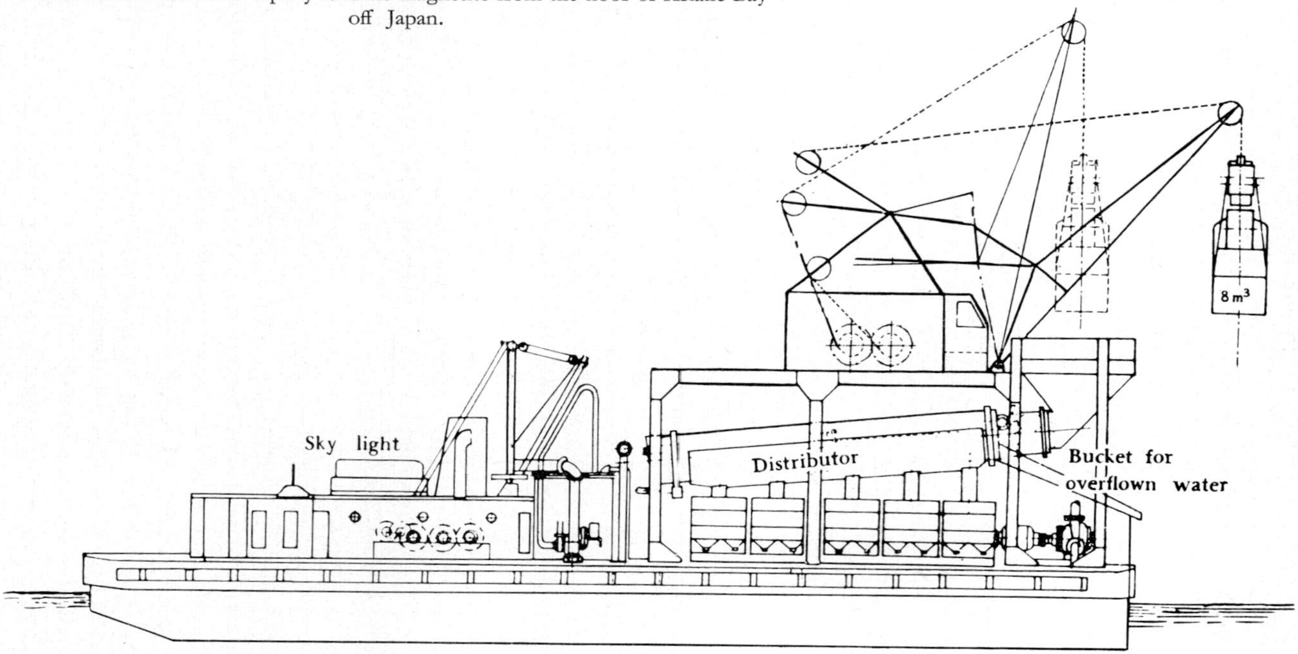

tons per hour are common with a single dredge and 30-ton buckets yielding production rates of 1,500 tons per hour have been designed (CRUICKSHANK, 1963). One version of a clamshell dredge bucket has been designed to mine gold-bearing sediments in the channels of southeastern Alaska at a depth of 800 ft. The jaws of this bucket are designed to be closed by pistons powered by hydrostatic pressure at this depth. At least one of these buckets has been constructed for the Williams Hydraulics Company of Oakland, California, but, as yet, has not been put into commercial production. Another clamshell dredge, designed for the mining of alluvial tin deposits in southeast Asia, has twin 6-cubic-yard buckets and is expected to attain a 308,000 cubic yard per month production rate in 135 ft. of water.

A great advantage of the clamshell dredge is its flexibility in operating in different depths of water with no major adjustment to the equipment. This type of dredge also incorporates a number of other advantages, such as being able to work in swells without loss of efficiency, ease of changing of bucket types as digging conditions require, and low maintenance. A clamshell dredge is presently used to mine magnetite from depths of about 130 ft. in Ariake Bay off Japan.

Bucket-ladder dredges

The digging mechanism of the bucket-ladder dredge consists of an endless chain to which buckets are attached. The chain is mounted on a truss or ladder and is drawn over the periphery of the ladder. A bucket-ladder dredge is shown in Fig. 69. The bucket-ladder method is a continuous process and production rates with these dredges normally average 150,000– 400,000 cubic yards of solids per month. Production rates depend on bucket size, power applied, and digging conditions. As the heavy ladder over which the chain and buckets are drawn can be rested on the bedrock or bottom during the dredging operation, this method of dredging allows excavation of moderately hard bedrock, a very important advantage in mining placer deposits in which the values are concentrated on or just within the surface of the bedrock.

The dredged material is drawn up the ladder, and, as the buckets

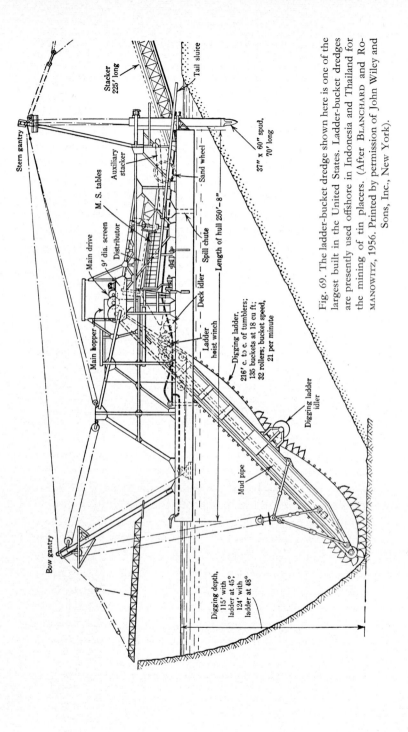

Fig. 69. The ladder-bucket dredge shown here is one of the largest built in the United States. Ladder-bucket dredges are presently used offshore in Indonesia and Thailand for the mining of tin placers. (After BLANCHARD and ROMANOWITZ, 1956. Printed by permission of John Wiley and Sons, Inc., New York).

ride over the rear-drive sprocket, they dump their load into a hopper from which the dredged material is fed to various screening and concentrating devices to separate the valuable minerals from the gangue. The gangue is normally dumped off the rear of the dredge as it proceeds forward through the deposit. Forward motion is accomplished by hauling in on headlines which are anchored several thousand feet, in front of the dredge. The anchors are moved as needed by utility boats. At sea, these dredges will normally ride out storms on the headlines.

The bucket-ladder dredge is limited to the depth at which it can work below sea level. Presently, these dredges are operating at depths of about 150 ft. below the water level. Because of a strength of materials problem in the chain and ladder, it is doubtful if such dredges can be built to work effectively in water depths greater than about 200 ft. The production costs attained by the ladder dredges operating offshore amount to U.S.$ 0.20–0.30 per cubic yard of material dredged. Bucket-ladder dredges are currently operating offshore in Thailand and Indonesia.

Hydraulic dredges

Until recently, the hydraulic dredge was used by mining companies mainly to remove overburden from ore deposits. The hydraulic dredge finds its greatest application in moving unconsolidated, low-specific gravity sediments over long distances in areas where a continuous supply of water is available. This dredge is not efficient in digging in consolidated materials, however, soft rocks can be dredged with it. The hydraulic dredge consists of a hull on which is mounted a suction pipe and support, pump with motors and controls, and a discharge line. Commonly, when digging in semi-consolidated sediments, a cutter head is mounted on the lower end of the suction pipe to agitate the semi-compacted sediments. A hydraulic dredge is shown in Fig. 70.

Normally, the hydraulic dredge is anchored by headlines set off to the front and sides of the dredge. The lines are hauled on, not only to hold the suction pipe against the sediment bank being mined, but to move the dredge laterally along the sediment bank. Hydraulic dredges are commonly used in dredging canals and in providing fill for the creation of land in near shore or low-lying areas. These dredges have also been used to mine sodium-sulphate deposits in Canada (ROMANOWITZ, 1962). In removing the overburden from iron-ore deposits at Steep Rock Lake in Canada, two 36-inch hydraulic dredges used 10,000 hp. on each of two pumps to move about 160-million cubic yards of glacial silt, gravel and boulders in addition to a lake overlying the iron-ore body, out of one valley and into another. These dredges worked for 5 years with discharge lines almost 4 miles

Fig. 70. Hydraulic dredges are not commonly used in the mining of metallic ore deposits, however, they are presently used in the mining of sand and gravel and shell deposits and in removing overburden from ore deposits. In isolated instances, they are used in the mining of tin placers. Hydraulic dredges have also been used in the mining of offshore magnetite deposits in Japan. The dredge shown in this photo, is the Hydro-Quebec, a 36-inch giant which can pump boulders 30 inches in diameter and weighing more than a ton. This dredge has a total of 10,000 hp. and can dig to a depth of 50 ft. below water level. (Photo courtesy of the Ellicott Machine Corporation, Baltimore, Maryland).

long and a static head of almost 600 ft. Such dredges may cost about U.S.$ 3.5 million to build and will remove unconsolidated gravels from water depths up to 200 ft. and move them at least 1 mile for a cost of about U.S.$ 0.25 per ton of solids.

Air-lift dredges

Although used in a variety of applications in lifting solids suspended in a fluid, the air-lift method of dredging solids had never been applied on a significant scale to a mining application until it was put to work dredging diamondiferous gravels from the sea floor off the west coast of Africa.

By injecting air into a submerged pipe at about 60% of the depth of submergence, the density of the fluid column inside the pipe can be lessened, forcing the fluid column to rise in the pipe line. If the top of the pipe is not too far above the surface of the water, the air-water mixture will overflow it. Water rushes into the bottom of the pipe to replace that lost in the overflow at the top and the capacity for lifting solids can be substantial. Air-lifts are about as uncomplicated as any device for moving materials can be. Many times they have been constructed on the site when a dredging device was needed immediately and conditions would allow the operation of such a device. The U.S. Navy used a field-built air-lift to dredge silt from the harbor in Subic Bay at a depth of 57 ft. A 2-inch pipe was used to deliver air at 75 lb. per square inch pressure into a 20-inch rise pipe. Held stationary on the bottom, this dredge was able to dig a hole 10 ft. in diameter and 3–5 ft. in depth in the silt. The same dredge was used in depths of 70 ft. and was able to lift the water–sediment mixture to a height of about 40 ft. above the sea surface. This author has seen small lead plates lifted from a flat surface and carried to the water surface with air-lifts.

In addition to their extreme simplicity (no submerged moving parts) air-lifts, because no suction is involved, can be used in conjunction with neutrally buoyant flexible pipe lines. The depth at which such devices can be used, however, is severely limited as the cost of supplying compressed air increases exponentially with the depth of dredging.

Mining minerals from the sub-seabed strata

Other than oil and gas, the only mineral commodity presently produced from the sub-seabed strata not involving an access drift driven into the ore deposit from a shore-side shaft is sulphur. As on land, subsea-floor salt-dome sulphur deposits are mined by the Frasch method. This method is described earlier in this volume.

Petroleum deposits are exploited by drilling wells into the oil-bearing formations below the level of any gas accumulations and letting the oil flow to the surface through pipe lines placed in these wells. Frequently, the pressure in the oil formation is sufficient to force the oil to the surface. If it is not, pumps may be installed in the wells to bring the oil to the surface. In the offshore areas, the oil may be either gathered in tanks supported by piling above sea level or piped ashore by sea-floor pipe lines. Many design studies have been made for sea-floor oil-storage tanks but as far as is known no large-scale storage facilities on the sea floor has yet been put into operation.

THE MINING OF SURFICIAL SEDIMENTS FROM THE OCEAN FLOOR

From an economic standpoint, probably the most important sea-floor mineral deposits known of, but not yet exploited, are the phosphorite nodules and the manganese nodules. Many different methods of mining sea-floor surficial sediments can be envisioned. Unmanned crawler-type units which would submerge, fill with nodules and surface; manned crawler-type bathyscaphes which would serve as the motive power to pull scraper units along the bottom; and large-tonnage submarines with manned spherical control chambers and flooded storage chambers into which the nodules would be gathered, displacing the water, were among some of the more practical suggestions.

For the immediate future, however, only two methods of bringing the nodules to the surface on a commercial scale seem to have merit, the presently used deep-sea drag dredge, and a deep-sea hydraulic dredge. A clamshell-type dredge was considered, and, if the nodules prove to exist in depth in the sea floor, such a dredging device might prove practical in depths less than about 2,000 ft. At present, the general consensus of opinion is that the nodules are only one layer thick on the surface of the ocean-floor sediments. To economically justify a dredging process in which a long time is spent in lowering and raising the dredge bucket, the bucket must carry large loads to the surface. If the nodules are only one layer thick, the clamshell-bucket would have to be of tremendous dimensions to gather a very large load. For example, to gather a 5-ton load at a 90% pickup efficiency in an area that had a nodule concentration of 2 lb. per square foot of sea floor, the clamshell-bucket would have to scrape about 5,500 square feet of the ocean floor. Such a bucket would be about 75 ft. square when opened; it would be very difficult to handle such a device at the surface for unloading, In the case of mining sea-floor phosphorite, however, with surface concentrations in the order of 30–50 lb. per square foot in water less than about 500 ft. deep, clamshell-dredging would be a potential economic recovery method.

The deep-sea drag dredge

Probably the least complicated method of mining sea-floor nodules would be some form of deep-sea drag-dredge. The equipment involved is simple, inexpensive, and has been used by oceanographers for almost a hundred years to recover sea-floor sediments from depths as great as 30,000 ft. The dredge bucket would be somewhat redesigned over present models especially as it will be used to skim only a thin surface layer of material from the sea floor in a specific area where the characteristics of the bottom sediments are known. Fig. 71 illustrates the general set-up of a drag-dredge operation. Fig. 67 shows one type of deep-sea drag-dredge bucket used to recover tonnage samples of nodules from the sea floor.

Any number of drag-dredge systems could be devised. The following described system illustrates the general method. This drag-dredging operation would require two floating vessels. The dredging ship, a 2,000-ton ocean-going tug-type vessel with an installed 2,000 hp. winch for dredging to 5,000 ft., should have excellent maneuverability. It would contain living quarters for the entire crew of the mining operation. The ship should be designed so that the main generators

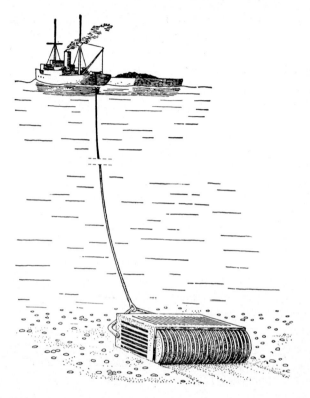

Fig. 71. The deep-sea drag dredge is a rather unsophisticated, but, in shallow water, effective, method of mining sea-floor surficial sediments. (After MERO, 1960b. Printed by permission of *Scientific American*).

can serve as a power source for the dredge-winch. During mining operations only a small fraction of the ship's power output would be needed for propulsion. Although it would be advantageous to have a ship designed for this mining operation, existing vessels could be modified to do the job at a cost of about U.S.$ 1 million for the modification.

The second vessel, an ocean-going barge, would be large enough to store about 5,000 tons of mined and cleaned nodules. This barge would carry equipment to separate the nodules from the debris recovered with the nodules and a small propulsion unit to enable it to follow the dredge ship during mining operations. It would have facilities for rapid transfer of mined material to a waiting transport vessel unless the barge were to be used as the transport vessel itself.

The dredge bucket should be as large as possible that will allow safe working conditions at sea. On a 2,000-ton ship, the largest bucket practicable would be about $20 \times 12 \times 3$ ft. The weight of this bucket would be about 3 tons and its load capacity, assuming a 65% filling efficiency and a 56 lb. per cubic foot bulk density of the dredged material, would be about 13 tons per haul. If 25% of the material dredged from the sea floor is assumed to be gangue such as sharks' teeth, whale ear bones, and fine sediments, this bucket should recover about 10 tons of nodules per haul.

The bucket will be allowed to drop at a free-fall velocity to the sea floor which, with proper hydrodynamic design of the dredge bucket, should be as much as 600 ft. per minute. A sonic pinger would be attached to the bucket so the operator would know when the bucket reached the sea floor. The bucket would be dragged across the sea floor until it was filled with nodules and then would be retrieved. Television cameras are presently available that could be mounted on the bucket to help guide filling of the bucket. The television cable could be embedded in the core of the dredging cable to avoid the complications of a two-cable dredge-line. The cost of this arrangement would be about U.S.$ 25,000 over that of the normal dredge cable and would certainly be justified by the increased efficiency in the operation it would allow.

As the bucket surfaces, it would be drawn up over a track on the back of the dredging ship and its load dumped into a hopper. From the hopper, the nodules would be fed into a pump and pipe line which would carry them to the storage barge. As the nodules are generally of a low density and a nonabrasive character, transport by rubber pipe line should involve no difficulties. On the storage barge the nodules would be separated from the gangue, crushed, and stored in hopper-type compartments designed to facilitate rapid transfer of the nodules to a transport vessel.

If a 300-day working year is assumed and the dredging ship and storage barge are chartered, the cost of running a deep-sea dredging operation can be estimated as follows:

(*A*) Fixed operational cost per year:

(*1*) Chartered dredging ship at U.S.$ 2,500 per day, 365 days per year: U.S.$ 912,500

(*2*) Storage barge, chartered at U.S.$ 500 per day, 365 days per year: 185,000

(*3*) Depreciation and maintenance on equipment (sonic pingers, dredge buckets, cameras, depth recorders, crushers, transfer system, etc: 320,000

(*4*) Labor, supervision of mining operation and overhead: 667,500

Total fixed operational costs per year U.S.$ 2,085,000

(*B*) Variable operational costs per year (largely dependent on the depth of dredging):

(*5*) Winch, at U.S.$ 5,200 per ton of capacity: Depreciation and maintenance at 40% per year of total cost;

(*6*) Cable, replaced four times a year at full cost;

(*7*) Power, figured at U.S.$ 0.02 per horsepower-hour, operating 300 days per year, 20 hours per day.

It is assumed that the mining operation will be shut down about 18% of the time due to bad weather, ore carrier delays, etc. In a 24-hour working day, 4 hours per day would be allowed, on the average, for equipment maintenance shutdowns. By using the above data and assuming an overall mechanical and electrical efficiency of the hoisting system of 65%, while maintaining a 2.5 safety factor on the cable, and assuming a bucket-dredging velocity of 3 ft. per second on the sea floor, data can be developed which indicate the production costs of mining deep-sea nodules from various depths of dredging. Such data are listed in Table XXXVII. The total weight being hoisted will be a combination of the submerged weight of the load and dredge bucket and the drag forces exerted on the bucket and cable. A more detailed discussion of these forces and the power required to overcome them can be found in a publication by MERO (1959).

The rate and cost of production in a deep-sea drag-dredge mining

TABLE XXXVII

NODULE PRODUCTION RATES AND COSTS USING THE DEEP-SEA DRAG DREDGE[1]

Depth of dredging (ft.)	Cable size (inches)	Lowering velocity (ft./min)	Raising velocity (ft./min)	Total drag (tons)	Steady state weight hoisted[2] (tons)	Average power (hp.)	Cycle time[3] (min)	Rate of production (tons/day)	Operating costs (U.S.$/year)	Cost of production (U.S.$/ton)
1,000	1	300	400	0.985	10.91	407	20.08	598	2,167,200	12.10
1,000	1	300	750	3.465	13.39	935	18.66	643	2,214,700	11.50
1,000	1	300	1,000	6.175	16.10	1,502	18.23	658	2,265,700	11.50
3,000	1	300	400	1.000	11.61	434	33.05	363	2,172,600	20.00
3,000	1	300	750	3.517	14.13	990	28.80	416	2,227,600	17.85
3,000	1.25	300	1,000	6.258	17.16	1,600	27.80	431	2,285,380	17.70
3,000	1	600	750	3.517	14.13	990	23.80	505	2,227,600	14.70
3,000	1.25	600	1,000	6.258	17.16	1,600	22.50	534	2,285,380	14.30
5,000	1	300	400	1.115	12.43	465	46.02	261	2,188,400	28.00
5,000	1.25	300	750	3.568	15.34	1,072	38.94	308	2,247,800	24.40
5,000	1	600	400	1.115	12.39	465	37.69	318	2,188,400	23.00
5,000	1.25	600	750	3.568	15.34	1,072	30.61	392	2,247,800	19.10
5,000	1.25	600	1,000	6.360	18.43	1,720	28.44	421	2,311,300	18.30
5,000	1.50	600	1,500	14.290	27.77	3,880	26.27	456	2,520,000	18.40
5,000	2	600	2,000	25.420	40.31	7,520	25.19	476	2,867,500	20.50
8,000	1.25	300	750	3.605	16.70	1,168	55.10	218	2,371,180	32.10
8,000	1.25	600	750	3.605	16.70	1,168	41.78	287	2,371,180	27.60
10,000	1.25	300	750	3.696	18.70	1,308	64.26	187	2,304,300	41.20
10,000	1.25	600	750	3.696	18.70	1,308	47.60	252	2,304,300	30.60

[1] With equipment and working conditions as outlined and assumptions of a nodule concentration on the sea floor of 1 lb. of nodules per square foot and a dredge pick-up efficiency of 80%.
[2] Sum of drag forces plus weight of bucket and load in sea water.
[3] Time required for complete dredge cycle including lowering, loading, raising, and unloading.

operation are very dependent on the depth of dredging, primarily because of the time it takes to raise and lower the dredge-bucket. Hoisting velocities of 2,000–3,000 ft. per minute are standard practice in the mining industry. The drag resistance of the dredge bucket in hoisting it through the water varies as the square of the velocity while the power required to overcome the drag varies as the cube of the velocity. Above about 1,000 ft. per minute the power to overcome the drag forces, if we assume no streamlining of the bucket, is so high that, even with the greater rate of production, the unit cost of production begins to rise again. This rise is reflected in the data of Table XXXVII where the 1,500-ft. per minute hoisting velocity from 5,000 ft. incurs a greater production cost than the 1,000-ft. per minute hoisting velocity. There is a maximum economic limit to the hoisting velocity for a given dredge bucket which can be determined by the methods of calculus. Streamlining of the dredge bucket would allow economic hoisting velocities in water greater than 1,000 ft. per min, but probably not exceeding 2,000 ft. per min.

In water less than about 5,000 ft. deep it might be possible to operate two dredge buckets off the same ship at the same time. The buckets could be attached to either end of the hoisting cable eliminating the need for a storage drum and second cable. While one bucket was ascending, the other would be descending. Such an operation would cut the unit production costs by a factor of two and would appreciably lower the capital investment in the hoisting system.

Because the time it takes to fill the dredge-bucket by dragging it along the bottom can be an appreciable part (over 70% in some cases) of the dredging cycle, concentrated deposits of the nodules can help decrease the production costs as shown in Table XXXVIII. There would be little detailed control over where the dredge bucket lands on the sea floor in this method of dredging; however, deposits of manganese nodules are generally so vast in lateral extent that random cuts through a deposit could be made for years without one cut ever crossing the path of another.

The capital cost of a drag-dredging system, as outlined here, can be expected to be about U.S.$ 2,000,000 if the ship and barge are chartered. This price would include modifications to the dredging

TABLE XXXVIII

DRAG DREDGE PRODUCTION RATES AND COSTS AT VARIOUS SEA-FLOOR NODULE CONCENTRATIONS[1]

Depth of dredging (ft.)	Lowering velocity (ft./min)	Raising velocity (ft./min)	Production rates (tons/day) at various nodule concentrations (lb./ft.²)				Production costs (U.S.$/ton) at various nodules concentrations (lb./ft.²)			
			0.5	2	5	10	0.5	2	5	10
1,000	300	750	396	935	1,157	1,460	18.60	7.90	6.40	5.05
3,000	300	750	298	521	614	655	24.90	14.25	12.10	11.30
3,000	600	750	349	667	825	900	21.30	11.10	9.00	8.25
5,000	300	750	238	362	404	436	31.50	20.70	18.60	17.20
5,000	600	1,000	300	530	626	666	25.70	14.55	12.30	11.55
8,000	600	750	225	333	369	383	35.10	23.70	21.40	20.60

[1] Other operating conditions are the same as for Table XXXVII.

ship and barge, the dredging system, and design and start-up costs. Exploration costs before the development of the mining equipment would be extra. The cost of a 2,000-ton ship to be used in this operation could be expected to be about U.S.$ 6,000,000 if built in an American yard. Because the underway demands on the ship in this operation would not be great, an older ship could be modified for the job. The barge needed in the operation outlined in this chapter with attendant nodule cleaning, crushing, and storage equipment could be purchased and outfitted for about U.S.$ 1,500,000. A used hull would be satisfactory for this type of operation.

The deep-sea hydraulic dredge

Any large-scale, efficient operation to mine sea-floor sediments with present technological means would require some form of a hydraulic dredge. Normally, hydraulic dredges operate with a motor and a pump inside the hull of a vessel. The pump is generally located near or just below the level of the water in which the vessel is floating. In order to draw water and dredged sediments through the pump, the pump must develop a vacuum at its suction end. Hydraulic dredges of this design are, consequently, severely limited to the depth from which they can pump. Operating in the atmosphere, the maximum suction they could develop is atmospheric pressure or about 34 ft. of water. Developing this vacuum, however, would cause the water at the suction end of the pump to boil. In general, hydraulic dredges work with an effective suction head of about 25 ft. of water. If we neglect fluid friction and assume a 10–1 fluid-solids weight ratio in the pipe line, the maximum depth from which such a dredge could lift sediments, assuming a uniform density of the water at all depths, would be about 250 ft. Of course, the fluid-solids ratio could be increased and the dredge would work in greater depths, however, power is being expended to overcome fluid friction as well as lift the solids in the pipe line and the point would soon be reached at which power costs far exceeded the value of the dredged material. Unless the depth of water is less than several hundred feet, therefore, the pump on any deep-sea hydraulic dredge must be submerged.

Although it must be submerged, the pump need not be operated near the sea floor. The factors controlling the location of the pump in regard to the surface of the ocean will be the fluid-solids ratio of the material in the pipe line and the fluid velocity at which the dredge is operated. It is advantageous from a number of standpoints to operate with the pump as close to the surface as possible and the following calculations, therefore, will assume such a position for the pump.

The major components of a deep-sea hydraulic dredge are a pipe line, pump and motor, suction heads, and a float. Although the weight of the dredge could be suspended from a floating vessel, such a system would be somewhat disadvantageous. Vertical oscillations of the vessel due to wave motions would be transmitted to the dredge introducing alternating stresses in the pipe line that might cause failure of the pipe line. Also, with such a system, the vessel and dredge would be extremely vulnerable in case of a sudden storm. The deep-sea hydraulic dredge, therefore, should be supported by floats which are an integral part of the dredge. As shown in Fig. 72, the main float tank is submerged below the turbulent surface layer of ocean water so no vertical wave motions will be transmitted to the dredge. A stabilization float is provided at the surface to keep the dredge afloat. As over 98% of the weight of the dredge is supported by the main float, no vertical oscillations of wave motions acting on the stabilization float would be transmitted to the dredge as a whole. The pump and motor of this system would be contained in the main float. The main float is over-sized and, thus, is ballasted with sea water to give the dredge proper trim. Should there be a failure in either the motor or pump, the ballast can be pumped out of the main float tank causing it to rise to the surface. A manhole at the top of the float tank would allow easy access by the crew to make repairs. In case of a sudden pump or motor failure, back flow valves along the pipe line would automatically open to vent the falling nodules and prevent them from jamming in the pipe line.

The calculations included here are for a dredge designed to operate in 10,000 ft. of water; however, the cost estimates are listed so that they can be extrapolated to other operating depths within the strength limitations of the materials used in the construction of the dredge.

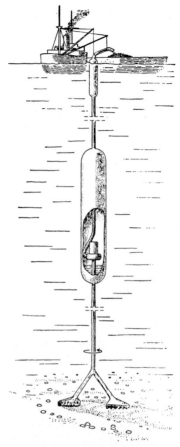

Fig. 72. The deep-sea hydraulic dredge is essentially a giant vacuum cleaner designed to gather a thin surficial layer of material, but disturbing the underlying sediments as little as possible. The particular dredge illustrated would rotate around its vertical axis so that the dredge heads would cover as large an area of the sea floor with as small a lateral motion of the dredge as possible. (After MERO, 1960b. Printed by permission of *Scientific American*).

The design considerations of the dredge are:
(*1*) Depth of operation: 10,000 ft.
(*2*) Maximum diameter of nodules to be dredged: 6 inches.
(*3*) Diameter of pipe line: 20 inches, inside diameter.
(*4*) Velocity of fluid flow in pipe line: 15 ft./sec.

(5) Fluid-solids ratio in pipe line: 10–1 by weight.
(6) Percent of nodules in fluid: 5% by weight (assuming 50% of the solids dredged will be gangue materials).
(7) Specific gravity of solids in fluid: 2.1.
(8) Foul weather downtime: 20% of the total time.
(9) Equipment downtime: 16% of the total time.

The weight of the dredge would be about 2,500 tons distributed as follows:

(1) Pipe line (1-inch wall): 10,000 ft. at 225 lb./ft.:   1,125 tons
(2) Pump: 20-inch diameter: 25 tons
(3) Motor: 8,000 hp.   30 tons
(4) Main float: 20-ft. diameter, 3-inch wall, 250-ft. long: 1,050 tons
(5) Stabilizing float: 6-ft. diameter, 2-inch wall, 30-ft. long: 10 tons
(6) Air in main float: pressurized to 450 lb./square inch: 100 tons
(7) Weight of fluid in pipe line not buoyed by sea water: 70 tons
(8) Miscellaneous: 100 tons

The buoyancy of the main float will be about 2,635 tons and that of the stabilizing float about 28 tons, therefore, the main float tank will be ballasted with about 150 tons of sea water.

The safety factor in the pipe line using an 80,000 lb. per square inch grade of steel will be about 3.0. If a greater safety factor is desired a pipe line with a tapered wall thickness can be used.

Fluid velocity necessary to carry the nodules in the pipe line

The law which governs the rate of fall of objects through a viscous medium at Reynolds' numbers greater than 0.5 is known as Newton's law and may be expressed in English units as:

$$V_t = 6.55 \left[\frac{P_s - P_w}{P_w} (d) \right]^{\frac{1}{2}}$$

Where,

V_t = the terminal velocity of the solid falling in static fluid under the action of gravity (ft./sec)
P_s = the specific gravity of the solid
P_w = the specific gravity of the fluid
d = the diameter of solid, assuming a spherical shape (ft.)

A number of experiments have been performed by dropping manganese nodules through a column of water (MERO, 1959). The experimental results involving spherical nodules closely agreed with the theoretical calculations. In actual practice when solids are being carried upward by a column of water, the velocity to support the solids is about 10% greater than the falling velocity of that solid in a column of still water. Other factors such as wall effects, hindering effects, and sphericity factors will modify and lessen the fluid velocity necessary to support the nodules.

Although the hydraulic dredge would be designed to mine a particular deposit of nodules, its application would be general in all deposits in which the diameter of the nodules to be mined did not exceed about half the diameter of the pipe line. The depth of the deposit and the size of the nodules are the prime design considerations. Fortunately, sea-floor photographs showing manganese nodules indicate that the nodules within many deposits have relatively uniform shapes and sizes (Fig. 51, 56, and 57).

The fluid velocity necessary to support a 6-inch diameter nodule of a specific gravity of 2.1 would be about 4.7 ft. per second, disregarding wall-effect and hindering factors. The average diameter of all nodules entering the pipeline would be less than 6 inches, thus, the average fluid-flow velocity necessary to carry all the nodules would be somewhat less than 4 7 ft. per second.

Power required to operate the dredge

The pump would be a standard, 20-inch, centrifugal dredge pump and would be operated with the pump axis in a vertical position as shown in Fig. 72. The suction opening of the pump would be connected to the upper end of the delivery pipe line. The power required to operate the dredge can be determined from standard fluid mechanics formulae.

The friction head, $h_L = f \dfrac{L}{D} \dfrac{V^2}{2g}$. In this case, $f = 0.0180$ as determined from Table XXXIX, and

$$h_L = \frac{(0.0180)(10{,}000 \text{ ft.})(15 \text{ ft.}/\text{sec})^2}{(20/12 \text{ ft.})(2)(32.2 \text{ ft.}/\text{sec}^2)} = 377 \text{ ft.}$$

TABLE XXXIX

VALUES OF "F" FOR USE IN THE WEISBACH EQUATION TO CALCULATE FRICTION HEAD LOSSES FOR FLUID FLOW IN STRAIGHT, SMOOTH PIPES

(After O'Brien et al., 1937)

Diameter of pipe (inches)	Mean velocity of fluid flow, V (ft./sec)		
	10.0	15.0	20.0
10	0.0206	0.0201	0.0197
12	0.0201	0.0196	0.0192
14	0.0196	0.0192	0.0188
16	0.0192	0.0188	0.0184
18	0.0188	0.0183	0.0181
20	0.0184	0.0180	0.0177

The density of the fluid being pumped will be about 71 lb. per cubic feet. The sea water will buoy $64/71$ (10,000 ft.) = 9,020 ft. of the fluid in the pipe line. Assuming the dredge must lift the nodules and fluid 20 ft. above sea level to get this material into the barge, the pump must lift a total of 1,000 ft. of the pipe-line fluid. The power required to overcome the weight of the fluid and the fluid friction will be:

$$E = \frac{p}{w} + \frac{V^2}{2g} + Z + h_L = 0 + \frac{(15 \text{ ft./sec})^2}{2(32.2 \text{ ft./sec}^2)} + 1{,}000 \text{ ft.} + 377 \text{ ft.}$$

= 1,381 ft.-lb./lb. of fluid; the weight of the material being lifted is $(15 \text{ ft./sec})(20/12 \text{ ft.})^2 \left[\frac{\pi}{4}\right] (71 \text{ lb./ft.}^3) = 2{,}320$ lb./sec; therefore,

the horsepower required is $\dfrac{(1{,}381 \text{ ft.-lb./lb.})(2{,}320 \text{ lb./sec})}{550 \text{ ft.-lb./sec/hp.}} = 5{,}830$ hp.

Assuming a 75% overall electrical, mechanical, and hydraulic efficiency of the system, the installed horsepower should be:

$$\frac{5{,}830 \text{ hp.}}{75\%} = 7{,}780 \text{ hp.}$$

Suction heads

The suction heads of a deep-sea hydraulic dredge as described herein can be of a number of different designs. A few of these designs could be:

(*1*) A simple drag head fanning out to a width of 15 ft. The suction mouth opening would be 15 × 0.5 ft., and the suction velocity would be about 5 ft. per second. Because no provision is made to propel this dredge through the water at relatively high rates of speed, nodule concentrations of over 10 lb. per square foot would be required for economical operation.

(*2*) A simple drag head fanning out to a width of about 15 ft. This drag head would be propelled back and forth across the ocean floor while the whole dredge moved very slowly (at velocities of less than 0.2 miles per hour) in a direction transverse to the sweeping action of the dredge head. This method would require a means of propulsion at the bottom to drive the suction head back and forth would complicate the design and operation of the dredge.

(*3*) Two dredge heads could extend down from an angled fitting. This arrangement is shown in Fig. 72. The dredge heads would be about 100 ft. apart at the sea floor. The whole dredge would be rotated slowly from the surface at a speed of about one revolution per minute. Mouths, 8-ft. wide, on each dredge arm would sweep over about 90 square feet of sea floor per second. Nodule concentrations as low as 2 lb. per square foot of sea floor could be mined by this system, assuming a 70% pickup efficiency. Propulsion devices could be mounted at intervals along the pipe line to cause the dredge to move laterally through the water to provide new ground for the suction heads. Ocean currents, however, may be sufficient to provide this lateral motion. The rate of travel of the dredge should be about 10 ft. per minute.

As manganese nodules are found lying loose at the surface of the soft sea-floor sediments, no mechanical device would be required at the sea floor to free the nodules. Pronged teeth, similar to the prongs on a hayrake, would extend below the dredge heads to lift the nodules out of the sediment and start them into the suction mouth of the dredge. The dredge suction mouth would ride above the sediment on skids to avoid removing too much of the sediments on which the nodules are resting. The largest nodule that could be taken up by the dredge in each of the three methods just described would be about 6 inches in diameter. Care must be taken in the design of the suction head mouth to be certain the flow of water through the sea-floor opening of the dredge is uniform and is of sufficient velocity to draw the nodules into the dredge mouth and to carry them into the pipe line.

The main advantage of the dredge described herein is that no major piece of electrical equipment is submerged other than the television cameras mounted on the dredge heads. Although the motor and pump would be operating about 1,000 ft. below the surface, they would be inside an air-pressurized float tank. A conventional motor could be used, and the pump would not require any radical additional design over conventional dredge pumps.

Television cameras would be mounted on the dredge heads to watch the mining operation at the sea floor and to give the operator information as to the efficiency of the dredge in cleaning the sea floor. As the dredge would be operating like a vacuum cleaner, drawing all loose sediments into the pipe line, no clouds of sediment would be raised to obscure the view of the sea floor. The operator would have gross control of the location of the dredge heads on the sea floor. The television cameras, thus, would be of prime importance in controlling the dredging operation so reasonably close parallel cuts can be made through a deposit without excessive overlap.

Production rate

Pumping at a rate of 2,320 lb. of fluid per second of which 5% is nodules would result in a production rate of nodules of about 4,180 tons per 20-hour working day, or 1,220,000 tons per 292-day working year. A dredge as shown in Fig. 72, rotating at one revolution per minute, having twin suction arms with 8-ft. suction mouths, operating about 100 ft. apart on the sea floor, could be used for this operation. Working in a nodule deposit with a sea-floor nodule concentration of about 2 lb. per square foot at a pick-up efficiency of about 70% this dredge would be able to make a 4,180-ton per day production rate without difficulty.

Capital and production costs

Production costs are calculated so that they can be applied to any depth of dredging with the hydraulic dredge. The operating costs listed in Tables XL and XLI include depreciation, overhauling, major repairs, painting, interest, taxes, insurance, and other incidental ownership expense.

Power will be taken for the dredge motor from the main propulsion motors of the ship. While dredging, very little power will be required for propulsion. A total of about 10,000 hp. should be available. If the ship's propulsion system cannot supply this much power, an auxiliary motor-generator must be installed. For pumping at a 75% overall efficiency, 0.778 hp. per foot of depth is needed. The cost of this power can be calculated thus:

(U.S.$ 0.02/hp.-hour)(5,840 hours/year)(0.778 hp./ft. of depth) = U.S.$ 91.00/ft. of depth/year.

Thus mining costs for deep-sea hydraulic dredging, assuming a production rate of 1.22 million tons per year, could be expected to be:

$$\frac{\text{U.S.\$ } 2,625,000}{1,220,000 \text{ tons}} + \frac{(\text{U.S.\$ } 79.70 + \text{U.S.\$ } 91.00)\, d}{1,220,000 \text{ tons}} =$$

U.S.$ $(2.15 + 0.00014\, d)$/ton,

where d is the depth of dredging in feet. From 10,000 ft., therefore, the production cost could be expected to be $2.15 + 0.00014(10,000)$ = U.S.$ 3.55 per ton of nodules recovered.

The rate of production for a deep-sea hydraulic dredge is a function of the size of the pipe line, velocity of fluid flow in the pipe line, and the fluid-solids ratio and is independent of the depth of dredging. The production costs for a given set of parameters, therefore, can be extrapolated to various depths of dredging to obtain an estimate of the production costs. Production costs, capital costs, and other statistics for dredging from various depths of water are listed in Table XLII. The statistics listed in Table XLII apply for a dredge with design parameters as herein outlined.

TABLE XL

DEEP-SEA HYDRAULIC DREDGE OPERATING COSTS THAT ARE INDEPENDENT OF THE DEPTH OF DREDGING

Item	Capital cost (U.S.$)	Operating cost (U.S.$/year)
(1) Storage barge, 15,000-ton used hull:	800,000	280,000
(2) Barge auxiliary equipment:	400,000	160,000
(3) Suction heads:	200,000	200,000
(4) TV cameras, cables, and controls:	50,000	25,000
(5) Miscellaneous equipment:	600,000	200,000
(6) Design:	300,000	—
(7) Overhead:	—	300,000
(8) Dredge construction costs	500,000	—
(9) Labor for mining operation:	—	365,000
(10) Chartered control ship, 3,000-ton:	—	1,095,000
(11) Modifications to control ship:	1,500,000	—
Total capital costs that are independent of the depth of dredging:	U.S.$ 4,350,000	—
Yearly operating costs:	—	U.S.$ 2,625,000

TABLE XLI

DEEP-SEA HYDRAULIC DREDGE OPERATING COSTS THAT ARE DEPENDENT ON THE DEPTH OF DREDGING

Item	Capital cost (U.S.$/ft. of depth)	Operating cost (U.S.$/year/ft. of depth)
(1) Pipe line:	40.00	20.00
(2) Floats:	20.00	10.00
(3) Pump:	5.00	5.00
(4) Motor ($80/hp.)(0.778 hp./ft.):	62.40	23.70
(5) Power cable ($20/ft.)(10% of depth)	2.00	1.00
(6) Instrumentation	10.00	10.00
(7) Miscellaneous	20.00	10.00
Capital costs dependent on the depth of dredging:	U.S.$ 159.40	—
Yearly operational costs dependent on the depth of dredging:	—	U.S.$ 79.70

TABLE XLII

NODULE PRODUCTION COST ESTIMATES BY HYDRAULIC DREDGING FROM VARIOUS DEPTHS

Depth (ft.)	Horsepower required (hp.)	Equipment[1] capital cost (U.S.$)	Depth-dependent equipment operating costs/year (U.S.$/year)	Power costs at U.S.$ 0.02/hp-hour (U.S.$/year)	Fixed operational costs/year (U.S.$/year)	Total operational costs/year (U.S.$/year)	Production cost[2] (U.S.$/ton of nodules)
1,000	778	4,509,400	79,700	91,000	2,625,000	2,795,700	2.29
2,000	1,560	4,668,800	159,400	182,000	2,625,000	2,966,400	2.43
3,000	2,330	4,828,200	239,000	273,000	2,625,000	3,137,100	2.57
4,000	3,110	4,987,600	318,800	364,000	2,625,000	3,307,800	2.71
5,000	3,890	5,147,000	398,500	455,000	2,625,000	3,478,500	2.85
6,000	4,670	5,306,400	478,200	546,000	2,625,000	3,649,200	2.99
7,000	5,450	5,465,800	557,900	637,000	2,625,000	3,819,900	3.12
8,000	6,220	5,625,200	637,600	728,000	2,625,000	3,990,600	3.27
9,000	7,000	5,784,600	717,300	819,000	2,625,000	4,161,300	3.41
10,000	7,780	5,944,000	797,000	910,000	2,625,000	4,332,000	3.55
12,000	9,340	6,262,800	956,400	1,092,000	2,625,000	4,673,400	3.83
14,000	10,890	6,581,600	1,115,800	1,274,000	2,625,000	5,014,800	4.11
15,000	11,670	6,741,000	1,195,500	1,365,000	2,625,000	5,185,500	4.25
16,000	12,450	6,900,400	1,275,200	1,456,000	2,625,000	5,356,200	4.40
18,000	14,000	7,219,200	1,434,600	1,638,000	2,625,000	5,697,600	4.66
20,000	15,560	7,538,000	1,594,000	1,820,000	2,625,000	6,039,000	4.95

[1] Assuming a chartered control ship.
[2] At a production rate of 1,220,000 tons per year.

Influence of weather on a mining operation

The weather encountered in the dredging area can have a major effect on the success of a mining operation. After an extensive exploration campaign, the area chosen for mining will most likely be determined on the basis of the depth of the water, the distance to the market, grade and concentration of the nodules, and the nodule environmental conditions. If the above conditions are satisfactory, the weather then becomes important and may be a controlling factor in turning down an otherwise favorable area. It is unlikely that mining will ever be undertaken in an area with weather similar to that of the North Atlantic. Ocean cross-currents must also be taken into account. Fortunately, however, the most favorable deposits, from a grade and concentration standpoint, thus far found are located near the equator in the eastern Pacific and in the central south Pacific, both areas noted for calm seas.

Part of the exploration of a deposit would be a study of the climate of an area. The surface equipment would be designed to withstand the worst storm that can be expected in the area. The hydraulic dredge will be designed to withstand any storm. As the major support of the dredge is below the turbulent surface layer of the ocean, the dredge should be left undisturbed by any storm.

Processing of the manganese nodules

After being mined and moved to a processing facility, the manganese nodules can be reduced to salable products by a number of methods (MERO, 1959). In general, most of the processes that have been developed to recover manganese from low-grade ores would be suitable, from a technical standpoint, in recovering manganese from the nodules. Other processes, such as the Sherritt–Gordon process, could be used to separate and recover the copper, nickel, and cobalt in the nodules. A significant discovery in process experiments on the nodules indicates, that various metals in the nodules are associated with the separate manganese and iron mineral phases of the nodules. Copper and nickel are apparently present in the manganese phase in the nodules, while the cobalt is present in the iron phase, presumably as replacements of the dominant ions in the crystallites of the nodules.

By controlling the leaching conditions, it is possible to preferentially leach the manganese from the iron, thus effecting a very simple and inexpensive separation of the bulk of the nickel from the cobalt in the nodules (MERO, 1963, U. S. Patent pending). Because of the chemical and physical similarity of the elements, their compounds, and their ions, nickel and cobalt are generally difficult and expensive to separate, especially when they are in solution together.

If presently available processes are used to reduce the manganese nodules to salable products, the estimate of the cost of processing them to obtain manganese as MnO_2 and cobalt, nickel, and copper as metals is about U.S.$ 25 per ton of raw nodules (MERO, 1959). As the gross recoverable value of the elements in the nodules varies between U.S.$ 40 and about U.S.$ 100 per ton of nodules, this indicated processing cost is not too high. In any case, special process systems will probably be developed to take advantage of the particular physical and chemical characteristics of the manganese nodules which systems should be more economical than the more conventional processes presently available.

CHAPTER VIII

SOME ECONOMIC AND LEGAL ASPECTS OF OCEAN MINING

Over the years, many prominent mineral economists have expressed alarm at the rate at which mankind is using up the mineral resources of the earth. Taken at any particular point in history, the statistics concerning our rates of consumption and the remaining reserves can, indeed, appear alarming. But looking at these statistics over a period of years, we see that, although our world per capita consumption has been increasing at a steady rate, the reserves of unexploited minerals have also either been increasing or at least holding their own. Economists' prognoses are invariably heavily biased by statistics covering past performances. Because it is most difficult to quantify, technological innovation generally is not taken into account in predicting future world commodity resources. Since technological innovation is rather new and generally uneven in its performance, it is understandable why no reliable statistics of past performance have been amassed which will allow extrapolation of its performance into the future.

Although technology has allowed us to discover completely covered mineral deposits and has allowed us to mine very low-grade ore bodies, we have had to pay for this technology. Thus, technological innovation becomes one more cost in the production of minerals. As long as research can provide production cost cutting results, it pays for itself. Like increasing the efficiency of a productive operation, however, you finally reach a point of diminishing returns, and funds allocated for research designed to provide technological innovation no longer produce economic results. Production costs, then, again must rise for it is becoming increasingly difficult to discover and exploit mineral deposits on the land.

Even with the development of technology such that the minerals industry in the United States has been able to achieve a remarkable four-fold increase in per man productive capacity in the past 20 years, Americans have increasingly looked to foreign nations for supplies of raw materials. Because of rising standards of living and/or populations in these underdeveloped nations, with the concomitant rise in per capita consumption of minerals, and because of political difficulties, these, once profitable, sources of minerals are beginning to become less reliable. Because of political risks, an ore body in a foreign, underdeveloped nation generally must be two to six times the grade of a deposit of similar minerals in the United States before most companies will invest in it.

There remain few areas on the earth, to which to turn for mineral resources. Northern North America is comparatively untouched as a source of common industrial minerals, although the more valuable minerals have been mined for years in this area. Because of the climate, however, this area presents no end of difficulties as far as mining is concerned, not the least of which is the problem of transportation.

One major region left which could serve as a source of many industrial minerals is the sea. Although it has served as a source of salt for as long a period as the land has produced minerals, the sea has been relatively little exploited in relation to its potential. The major reasons for this situation are, I believe, a lack of information concerning what is in the ocean in the way of mineral resources, the relative richness of many of the mineral deposits therein, and the comparative ease and efficiency with which many of these deposits can be exploited. Although the problems connected with developing the mineral resources of the ocean may seem formidable to many, the technology involved is certainly an order of magnitude less formidable than putting men in orbit around the earth and about three orders of magnitude less expensive.

Studies concerning the technical and economic aspects of mining and processing sea-floor sediments were initiated at the Institute of Marine Resources of the University of California early in 1958. Conclusions from these studies indicate that many of the industrially important metals can be produced at 50–75% of the present cost of producing these metals from land deposits. Later studies by industrial concerns essentially confirmed the results of the University of California studies. These industrial studies cost several hundred thousands of dollars and are definitive enough to encourage the additional spending of millions of dollars to develop equipment and methods with which to mine ocean-floor deposits.

Eventually, political and population pressures will force the more highly industrialized nations into recovering many materials from the sea. What I would like to emphasize at this point is that substantial engineering data and calculations show that it would be profitable to mine materials such as phosphate, nickel, copper, cobalt, and even manganese at today's (1964) costs and prices. And I firmly believe that within the next generation, the sea will be a major source of, not only those metals, but of molybdenum, vanadium, lead, zinc, titanium, aluminum, zirconium, and several other metals as well.

In the ocean, nature is working on a truly grand scale to separate and concentrate many of the elements that enter sea water. The minerals that are formed are frequently found in high concentrations on the sea floor as in the pelagic, or far out, areas of the ocean there is relatively little clastic material deposited to dilute the chemical precipitates. Eventually, with developing technology and ever cheapening power costs, the common rocks may serve as sources of all the minerals society will need. Fig. 73 illustrates one of the reasons why the sediments of the ocean will likely be used first. The pelagic sediments contain an average of about ten times the amount of the industrially important metals as do the igneous rocks on land. These ocean-floor sediments have other advantages when being considered as a material to mine, that of being politically-free and royalty-free materials, they are widely distributed near most markets and are available to all nations on an equal basis. In addition, these materials are fine-grained, unconsolidated, and in a water atmosphere which makes the use of an automated hydraulic system for recovery practical.

In spite of these advantages, there are many reasons why there is no particular pressure to develop marine mineral resources on a greater scale. The present supply and demand situation concerning most mineral commodities is well in balance and is likely to remain so for

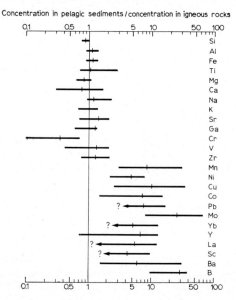

Fig. 73. Ratios of average elemental abundances in the Pacific pelagic sediments and igneous rocks. Ranges are shown by the lengths of the horizontal lines, and modes by the short vertical lines. (After GOLDBERG and ARRHENIUS, 1958).

some time. Although the reserves of many important industrial mineral commodities are quoted as being 30 or 40 years, the situation was the same 30 or 40 years ago. When the apparent reserves of any mineral commodity appear to be running out, the price of the commodity naturally increases, assuming a continuing demand. And enterprising minerals companies always manage to find another 20 or 30 years reserves, either through the discovery of new deposits or through the extension of the reserves definition to lower-grade deposits through increases in mining or processing efficiencies or even through price increases in the commodity.

Because of the great strides being made by scientists and engineers in increasing the efficiency of using land resources, it is unlikely that we would produce minerals from the sea because of a lack of them on the continents. Rather, we would produce minerals from the sea only because it would be less expensive than it is to produce these minerals from land sources. Of course, influences external to the supply and demand process, such as political climates, tax situations, economic systems which ignore market economics, pressures to preserve willderness areas for recreational purposes, and land pressures for living spaces, can and do change the atmosphere for exploiting various sources of various minerals.

THE MANGANESE NODULES AS A MINERAL RESOURCE

Probably the most interesting of the mineral deposits of the sea are the manganese nodules. They were apparently first recognized as a potential mineral resource by MERO (1952). Whether the sea-floor manganese nodules can be considered to be an ore in the sense of being economic to mine, process, and market in competition with the present sources of the metals in the nodules remains to be seen. Undoubtedly, it would be very profitable to mine certain of these deposits, even at present costs and prices. Calculations and laboratory experiments indicate that there should be no major problems in adapting existing industrial equipment and processes to the mining and processing of the manganese nodules. Indications from studies made in the past two years are, that the nodules should be ores of manganese, nickel, cobalt, and copper. Other materials such as molybdenum, lead, zinc, zirconium, several of the rare earth elements, and, possibly, iron, alumina, titanium, magnesium, and vanadium could be recovered as by-products.

If the manganese nodules prove to be economic to mine for the various metals contained therein, Table XLIII shows the statistics concerning the reserves that would be available from the nodules now speculated to be at the surface of the sediments of the Pacific Ocean. Assuming that only 10% of the nodule deposits prove economic to mine, it can be seen that there are, in general, sufficient supplies of many metals in these sea-floor deposits to last for thousands of years at our present rates of consumption.

Even assuming a world population of 20 billion people consuming metals at a rate equal to that in the United States at the present time, the reserves of most of the industrially important metals would still

TABLE XLIII

RESERVES OF METALS IN MANGANESE NODULES OF THE PACIFIC OCEAN

Element	Amount of element in nodules (billions of tons)[1]	Reserves in nodules at consumption rate of 1960 (years)[2]	Approximate world land reserves of element (years)[3]	Ratio of reserves in nodules/reserves on land	U.S. rate of consumption of element in 1960 (millions of tons/year)[4]	Rate of accumulation of element in nodules (millions of tons/year)	Ratio of rate of accumulation/rate of U.S. consumption	Ratio of world consumption/U.S. consumption
Mg	25.0	600,000	L[6]	—	0.04	0.18	4.5	2.5
Al	43.0	20,000	100	200	2.0	0.30	0.15	2.0
Ti	9.9	2,000,000	L	—	0.30	0.069	0.23	4.0
V	0.8	400,000	L	—	0.002	0.0056	2.8	4.0
Mn	358.0	400,000	100	4,000	0.8	2.5	3.0	8.0
Fe	207.0	2,000	500[5]	4	100.0	1.4	0.01	2.5
Co	5.2	200,000	40	5,000	0.008	0.036	4.5	2.0
Ni	14.7	150,000	100	1,500	0.11	0.102	1.0	3.0
Cu	7.9	6,000	40	150	1.2	0.055	0.05	4.0
Zn	0.7	1,000	100	10	0.9	0.0048	0.005	3.5
Ga	0.015	150,000	—	—	0.0001	0.0001	1.0	—
Zr	0.93	100,000	100	1,000	0.0013	0.0065	5.0	—
Mo	0.77	30,000	500	60	0.025	0.0054	0.2	2.0
Ag	0.001	100	100	1	0.006	0.00003	0.005	—
Pb	1.3	1,000	40	50	1.0	0.009	0.0009	2.5

[1] All tonnages in metric units.
[2] Amount available in the nodules divided by the consumption rate.
[3] Calculated as the element in metric tons. (U. S. Bureau of Mines Staff, 1956).
[4] Calculated as the element in metric tons.
[5] Including deposits of iron that are at present considered marginal.
[6] Present reserves so large as to be essentially unlimited at present rates of consumption.

be measured in terms of thousands of years. Development of the means to mine the manganese nodules, thus, could serve to remove one of the historic causes of war between nations, supplies of raw materials for expanding populations. Of course it might produce the opposite effect also, that of fomenting inane squabbles over who owns which areas of the ocean floor and who is to collect the protection money from the mining companies.

The results of another interesting set of calculations are shown in the last three columns of Table XLIII. By dividing the rate at which various metals are agglomerating in the manganese nodules now forming on the Pacific Ocean floor by the rate at which we are presently consuming these elements, it is seen that, in many cases, the elements are accumulating in the nodules faster than we could consume them, in fact, three times as fast in the case of manganese, four times as fast in the case of cobalt, as fast in the case of nickel, and so on. Once these nodules are being mined, therefore, the minerals industry would be faced with the very interesting situation of working deposits that grow faster then they can be mined.

Because of the general widespread availability of certain metals in the nodules, the reserve calculations shown in Table XLIII for those metals, of course, are of academic interest only; titanium and vanadium, for example. Also, there would be marketing problems in disposing of all the products, if the nodules were mined at a rate to produce the economically most significant metal in the nodules, which is nickel. For example, if an operation were designed to mine an average grade of the nodules to produce 100% of the United States consumption of nickel, that operation would also produce about 300% of its annual consumption of manganese, about 200% of that of cobalt, about 100% of that of titanium, about 300% of that of vanadium, about 500% of that of zirconium, and so on. Fortunately, the composition of the nodules varies markedly from location to location on the ocean floor, and a mining site can probably be chosen to yield nodules of a composition that allows disposition of all the metals and products without disrupting the market as far as the total amount of these materials consumed is concerned and without overproducing any one product.

ADVANTAGES OF DEEP-SEA MINING

Ocean mining offers many advantages which are not possible with traditional land mining. We have, in the ocean, materials that are available without removing any overburden, without the use of explosives, and without expensive drilling operations for sampling and ore breakage. With cameras, the complete deposit can be explored prior to mining—every ton of ore can be directly accounted for before mining starts. There will be no drifts to drive, shafts to sink, or town sites to construct in developing a deep-sea mine.

An ocean-mining operation, because it would be a whole new concept in mining, can be designed for automation from the beginning, which would result in new equipment designs not bound by traditions. The equipment would be very flexible to move from one area to another for the various types of nodules as the market demands. Sea transportation can be used to carry the mined material to most of the world's markets with no other form of transportation involved. About 75% of the material, and more in some cases, being mined and handled is salable in contrast to the 2% or so of today's copper and nickel ores. The grade and physical characteristics of the deposits are highly uniform over large areas. The non-abrasive character and low density of the nodules would allow the use of hydraulic systems for the transfering of the nodules throughout the mining and processing operations.

Political difficulties should be at a minimum, at least in the beginning, for deep-sea mining companies.

The unlimited amount of the sea-floor minerals should establish a base price and supply for nodule contained metals which certain of these commodities need, especially copper and cobalt. But most important, the sea-floor nodules should prove to be a less expensive source of manganese, nickel, cobalt, copper, and possibly other metals than are our present land sources.

LEGAL PROBLEMS INVOLVED IN OCEAN MINING

Marine beach deposits clearly lie within the jurisdiction of the coastal nation in which the beach is located. Normally there should be no dispute among various nations concerning the jurisdiction involved, save for the determination of the boundary between coastal states and the extension of those boundaries seaward. The seaward path of such extensions are important matters if they traverse offshore mineral producing beaches or oil fields. Various methods have been proposed and used in determining these dividing lines such as a seaward extension of the territorial boundary on land or a line extending seaward that is perpendicular to the trend of the shore-line and intersects it at the point where the land boundaries of the nations cut the coastal line. Unless the boundaries are fixed before commercial mineral deposits are discovered, we can expect problems to arise concerning ownership of offshore areas contiguous to two or more nations. A substantial legal battle is developing concerning the fixing of national boundaries in the North Sea as regards ownership of the potential oil and gas producing sediments of that body of water. Submerged beaches extend seaward for many miles off certain coasts and in a number of locations such as off the west coast of Africa valuable mineral deposits are contained therein. In addition to dividing boundaries, there is the problem of determining how far to sea a state's authority extends concerning sea-floor mineral deposits.

In the case of deposits lying inside the 3-mile limit, there exists a body of law through which disputes can be handled. This body of law concerning coastal waters, which is found in codes or traditions of a nation, has been formulated largely to regulate transportation or fishing activities in offshore areas. In general, the coastal nation has jurisdiction over the water to a distance of 3 miles from its coasts where use of these waters for the movement of ships is concerned. As long as legitimate commerce is the object of sailing in such waters no one gets particularily excited about shipping activities within the 3-mile limit. Outside the 3-mile limit is another matter.

Fishing rights have produced no end of disputes. Serious disputes concerning the right to fish up to the 3-mile limit have taken place in the past few years between Iceland and Great Britain, France and Brazil, and the United States and Ecuador. The value of offshore fisheries is becoming increasingly apparent to emerging nations which, in the past, have largely ignored such resources. Because no formal outer limit of control was ever agreed upon by the nations of the world, several nations have seen fit to extend their borders seaward by at least 200 nautical miles. Foreign vessels have been impounded by Ecuador for fishing within these limits and the fishermen fined.

Concerning minerals outside the 3-mile limit, President Truman, in 1945, solved some of the problems by proclaiming United States' ownership of all such resources in and on what can be construed as the continental shelf. No claim of authority over the waters overlying the continental shelf, beyond the 3-mile limit, however, was made. Generally, such claims are recognized by the community of nations if the proclaiming nation has the power to protect its claims.

Although no international dispute has arisen between the United States and other nations concerning offshore mineral deposits, a major dispute arose between the Federal Government of the United States and the various states off whose shores petroleum was being produced. Although oil had been produced from offshore deposits in California since before 1900, the total value of the petroleum produced between the years 1900 and 1940 was too small to fight about. After 1940, and especially after the end of the Second World War, exploration for oil in offshore areas expanded considerably and substantial quantities of petroleum were found. When the potential size of the revenues to be derived from leases, taxes, etc., became evident, legal disputes between the Federal Government and the various states arose.

As California was the first state to receive revenues from the production of offshore oil, it was the first to be sued. In June of 1947, the United States Supreme Court held that, as protection and control of the 3-mile belt of coastal waters has been and is the function of national sovereignty, that the Federal Government has paramount rights and power over the belt, and, incident to which is, full dominion over the resources of the soil under that water area, including oil-bearing rocks.

Even after this judgement was presented against the State of California, the State of Louisiana continued to issue leases on its offshore area. By the end of 1948, Louisiana had leased over 2 million acres of the floor of the Gulf of Mexico, many of these leases covering ground outside the 3-mile limit. At that time, the State of Louisiana claimed jurisdiction over the sea floor to a distance of about 27 miles from its coast. Late in 1948, a suit was brought by the Federal Government against the State of Louisiana and again, in June, 1950, the Supreme Court ruled for the Federal Government adding that, "If, as we held in California's case, the 3-mile belt is in the domain of the Nation rather than that of the separate states, it follows a fortiori that the ocean beyond that limit also is." A similar judgement was issued against the State of Texas in the same month (THOMASSON, 1958).

Following these decisions of the Supreme Court, practically all oil and gas operations in the continental shelf area came to a halt save for production from wells already drilled. As far as the operators were concerned the legal status of these areas was still in question. To clarify the status of the deposits in the offshore areas and to allow the oil companies to operate on the leases they had taken, Congress passed the *Submerged Lands Act* (Public Law 31, 83rd Congress, 1st Session, 67, Statute 29). The net effect of this act was to grant the coastal state jurisdiction over the submerged lands of the continental shelf and ownership of all minerals therein within the "territorial sea" i.e., within the 3-mile limit.

After the enactment of the *Submerged Lands Act*, Congress provided for the jurisdiction of the continental shelf outside the territorial sea area with the *Outer Continental Shelf Lands Act* (Public Law 212, 83rd Congress, 1st Session, 67, Statute 462). This act established the basic policy of the United States in regard to the ownership of the sea bed of the continental shelf outside the 3-mile limit as follows:

"Section 3a. It is hereby declared to be the policy of the United States that the subsoil and sea bed of the outer Continental Shelf appertain to the United States and are subject to its jurisdiction control, and power of disposition as provided in this Act." The term "outer Continental Shelf" is defined in this act as "all submerged lands lying seaward and outside of the area of lands beneath navigable waters as defined in section 2 of the *Submerged Lands Act* . . ., and of which the subsoil and sea bed appertain to the United States and are subject to its jurisdiction and control." This act also provided the framework for the issuance of mineral leases in the outer continental shelf and for the supervision and regulation of operations conducted pursuant to such leases.

These two acts did much to resolve the conflict concerning jurisdiction over the continental shelves contiguous to the United States and the mineral deposits on and therein, and provided a legal foundation on which private industry could lease, develop, and produce minerals from this area. The acts, of course, placed the jurisdiction of the United States, in regard to the production of minerals from the offshore area, considerably beyond any limit previously proclaimed or accepted by any nation. The continental shelf extends at least 250 miles seaward in places along the east coast of the United States.

If the precedent of the United States is followed, and we can expect that it at least will be followed if not exceeded in territorial claims, the legal status of the beaches and continental shelves is fairly well covered. It is under the jurisdiction of the coastal nation. As for sea water serving as a source of minerals, there is so much of it in relation to the amount being processed, no one cares. In any case, the water that will be processed will most likely always come from the territorial sea area.

Law pertaining to deposits of the deep sea

To the deposits of the deep sea, another, entirely different set of laws, apparently applies, the *law of the sea*. Although the deposits of the deep sea, such as manganese nodules, are enormous and widespread, they differ in composition from place to place and the deposits in one area can be much more favorable to mine, from an economic standpoint, than in another. There is some point, therefore, in having a body of law available and an enforcement agency to regulate exploitation of these mineral deposits. Probably the most interesting discussion concerning the law of the sea applicable to the mineral deposits of the deep sea is in a paper, presented before the American Chemical Society in Los Angeles, in April, 1963, by Mr. Wilbert M. Chapman. The following discussion is mainly abstracted from his talk.

At present, the law of the sea is a conglomerate of the practices of nations, decisions of international tribunals, treaty, law, World Court decisions, legal interpretations by "experts" etc., that has grown up piecemeal in the past 400 years as the nations began to indulge in maritime commerce (CHAPMAN, 1963). Chapman goes on to say, "Its main elements were hacked out by gunfire and sword in the approximate 200 years between the accession of the first Elizabeth to the throne of England and the conclusion of the Napoleonic Wars. Their prime function was to nail down the right of all vessels to use the high seas for peaceful commerce freely. With very few exceptions, this right has not been seriously challenged since the first decade of the 19th century. There is no right in international law which the nations are so quick and certain to war over to protect as this one because in these 400 years, the highways of the seas have become the vital arteries carrying the stuff of life among nations without which most of them could poorly support modern economic and political standards."

"In the classic period of the law from the end of the Napoleonic Wars to the end of the Second World War, the right of the fishermen of all nations to fish where they wished upon the high seas, without interference from any save their own sovereign, was carved into the law of the sea as deeply as the freedom of navigation on the high seas through a series of international arbitrations, treaties, and other acts."

As new technologies developed, their use of the sea was recognized by the law in granting the nations the right to lay submarine cable, and sometime afterward, pipe lines, on the bottom of the sea. In its time, flying over the seas became a right. Today, then, the four freedoms which the law of the sea seeks to protect are: (*1*) freedom of navigation; (*2*) freedom of fishing; (*3*) freedom to lay submarine cables and pipe lines; (*4*) freedom to fly over the high seas. By "high seas" is meant that part of the ocean not included in the territorial waters or in the internal waters of a state. Since the 16th century, there has been general agreement among the nations that between its shoreline and the high seas there was a narrow belt of the sea in which all nations had certain rights, connected with navigation, but which was the sovereign territory of the coastal state, and the waters and resources of the area were subject to the coastal state's exclusive ownership. This belt is termed the territorial sea. There was never absolute concurrence among the nations as to what the width of this territorial sea should be. From the early 19th century up until 1945, the breadth of the territorial sea, de facto, was 3 marine miles, mainly because those countries carrying about 90% of the the maritime cargo and having approximately the same proportion of the world's naval fire power said it was so.

The prime factor in upsetting this doctrine was an advance in technology which allowed companies to drill for oil on the continental shelf from platforms not connected with the land. At the time such drilling ensued, there was no legal basis in the U.S.A. for either the Federal Government or the states to claim ownership of these offshore oil deposits. Probably the strongest and oldest concept of the United States policy concerning this area of the sea was proclaimed by Thomas Jefferson, when he was Secretary of State, and that was that the breadth of the territorial sea was 3 marine miles and all outside that was res nullius, belonging to nobody, or, if you like, res communis, belonging to all.

In an attempt to forestall controversies which might arise between nations and not be so "easily" settled as was the fight between the Federal Government and the states, the Department of State of the United States inquired among the principal maritime nations as to whether they were agreeable to a modification of the customary international law so as to provide to the coastal state exclusive rights to the exploitation of the resources in and on the continental shelf contiguous to its coast. Naturally, they were agreeable, and accordingly, President Truman proclaimed, in September, 1945, that ownership of mineral resources in and on what could be construed as the continental shelf by the contiguous nation was the policy of the United States. No substantial controversy arose concerning this new principle and the United States thus initiated work in the United Nations which would lead to the General Assembly adopting this new principal as a tenet of international law.

Somewhere along the line, however, the fisheries people saw the enactment of such a law as an infringement on what they considered their rights. A great deal of bitterness was developed over this issue and the noise that the fisheries people created attracted a number of other groups who saw in an extension of the territorial sea, some threat to their rights. Admiralties of the world became worried over being denied passage in a few narrow straits previously open to all ships. The Arab nations saw in it a chance to restrict maritime trade to Israel and several other nations saw chances to restrict the trade of their economic rivals.

It was in this atmosphere that some 88 nations came together in Geneva, in the spring of 1958, to formulate and codify the Law of the Sea. Remarkably, a codification of the law was accomplished and the 1958 Conference on the Law of the Sea adopted four documents: (*1*) convention on the territorial sea and the contiguous zone; (*2*) convention on the high seas; (*3*) convention on fishing and conservation of the living resources of the high seas; and, (*4*) convention on the continental shelf.

The first convention defines the territorial sea in all aspects, except breadth, and sets down the rules under which nations shall conduct themselves in it. The second convention defines the high seas, declares them open to all nations, sets out specifically the four freedoms already mentioned and sets down the rules as to how nations shall conduct themselves on the high seas. It makes no mention of the sea floor except to define the rights and obligations of nations in regard to pipe lines or submarine cables on or in the bottom.

The third convention is different from the other three in that it formulates new law rather than merely codifying existing unwritten laws. In its first article it provides for: (*1*) the right for all nations to engage in fishing on the high seas subject to, (*a*) treaty obligations, (*b*) the interests and rights of the coastal states as provided for in the convention, and (*c*) the provision in the articles concerning conservation of resources, and (*2*) the duty for all nations to see that their subjects do what is necessary for the conservation of the living resources of the high seas. In its second article, the third convention defines conservation as the aggregate of the measures rendering possible the optimum sustainable yield from those resources, so as to secure a maximum supply of food and other marine resources, an eminently commonsensical statement. The next few articles provide means by which conservation can be demanded and obtained by either a coastal or a fishing state and the remaining articles show a creative feature of lawmaking by setting up the procedure by which disputes which may arise under any of the articles, and which cannot be settled among the disputants by peaceful means, can be settled by a special commission which can be convened as needed and which shall examine the dispute and make findings on the basis of "facts" in accordance with criteria laid down in the convention. The rulings of the commission are binding upon the disputants.

The fourth convention concerns the continental shelf. The chief feature of this convention is that it conveys to the contiguous coastal state two sovereign rights on the continental shelf: (*1*) that of exploring it, and (*2*) that of exploiting the natural resources found therein, exclusively. In exercising these rights, the coastal nation must not interfere unduly with other nations' rights in these areas pertaining to the superadjacent waters, principally navigation rights. Exercise of these rights also must not unduly impede the laying or maintenance of submarine cable or pipe lines in the area or with the conservation of the living resources of the sea, or with fundamental oceanographic or other scientific research carried out with the intention of open publication by other nations. It also must reasonably provide for safety of navigation around devices used by it in exploration or exploitation.

The major fault of this convention lay in its definition of the continental shelf. Normally, one thinks of the continental shelf as a geomorphic feature and subject to definition by competent marine geologists. Not so, however, at Geneva, for it turned out that there were at least three kinds of continental shelves in the minds of the delegates and these three were far from being similar either in their geologic form or outer boundary. There was the natural continental shelf defined by marine geologists and reflecting a tangible association with reality. This continental shelf was not of overwhelming interest to most of the delegates. There was the juridical continental shelf which existed off the coasts of certain countries by governmental proclamation. Examples of such shelves can be seen off the coasts of Chile, Ecuador, and Peru. Off these countries, there exists little of a shelf of the first kind and rather than be left without anything while great gifts of the sea floor were being made to other nations, these states proclaimed sovereignty over the sea, its bottom, and the resources in an on each to a minimum distance of 200 marine miles, presumably leaving the outer edge flexible for later adjustments as required. These were solemn acts of sovereigns and could not be retracted by them with dignity, nor could they be forced aside by international edict.

Then there is the political continental shelf. If a two thirds vote of duly accredited delegates to a United Nations Conference on the Law of the Sea is cast in favor of the definition of the continental shelf being so many miles wide or extending to a depth so many meters deep, that is the definition which will apply in international law. This definition of the third type of shelf is the one used in international law at this point. In it the shelf is defined as, "the sea bed and subsoil of the submarine areas adjacent to the coast but outside the area of the territorial sea, to a depth of 200 m (656 ft.) or, beyond that limit, to where the depth of the superjacent waters admits of the exploitation of the natural resources of the said areas." The words are so clear as to allow little dispute. Applicable situations, however, can be quite interesting. In the case of phosphorite nodules which are found on the shelves and the continental slopes to depths exceeding 10,000 ft., there is little doubt that the adjacent coastal nations own them. And the United States has already acted in the case of this material by leasing to the Collier Carbon and Chemical Company some 30,000 acres of Forty Mile Bank, an area which lies about 40 miles due west of San Diego and is separated from the coast of California by a basin exceeding 1,000 fathoms in depth.

Presumably also, the continental shelf can be extended by the coastal nation out over the edge of the geologic shelf, down the continental slope and on out over the deep ocean floor to whatever point a commercial minerals dredge can operate. Such an extension was probably not the intention of the Conference, but the wording of the articles of the Fourth Convention is so clear as to leave little room for maneuvering on this point. Such an interpretation of this article can be either advantageous or disadvantageous to the ocean-floor miner. It provides a mechanism by which some agency can control the deposits, issue leases on them and prevent interlopers from either exploring or mining the offshore sediments. The large mining companies would appreciate such control over the deposits for it would serve to protect the capital they invest in finding and exploring a deposit for which, of course, they would be expected to pay a fee. The legal corner stone of the present mining industry is that one company or man can gain exclusive control over a specific mineral deposit to develop and mine it as he sees fit. No such control over the mineral deposits of the pelagic ocean floor is a major deterent to the participation of the logical group to initiate exploitation of these deposits, the mining industry. The major disadvantage of interpreting this article in this fashion is the fee that would have to be paid to some nation for the privilege of mining the deposit and the time that would be lost in red tape activities. In addition, of course, nations without coastal shores or with shores not having favorable deposits nearby or having shores, the extension of the boundaries of which are inhibited by other nations lying between them and the open ocean as China, will raise a row about being factored out of ocean mining or having to pay some other nation for mining rights on the high seas. Such protest would only lead to interminable discussions in some pleasant areas of the earth, such as Switzerland, by the various politicians and diplomats who discuss such matters.

Thus, it would appear to be of greater advantage to the ocean miner that the convention governing the high seas take control at the base of the continental slope. Because of the large capital requirements for mining sediments of the deep sea, it is assumed that only sophisticated groups will be involved and that they should be able to settle any of their disputes among themselves. In case disputes cannot be peacefully settled, the convention regarding fisheries on the high seas offers a means of settling them, namely the convening of a commission to examine the facts of the dispute and to render a binding settlement. The convention regarding the fisheries on the high seas codifies other laws which should also be of great interest to the ocean miner.

Although the fisheries convention has no direct bearing on the legal problems in harvesting minerals of the deep sea, its indirect bearing on that subject is strong. There are strong resemblences between a fisheries operation on the high seas and a mining operation. Both operations represent high capital investments in recovery machines which are mobile and which, for purposes of maritime law, can be considered as merchant ships. The fundamental difference between the fishing operation and the mining operation is, of course, in the nature of the object of the quest. The fishing vessel is after fugacious objects while the mining vessel is after objects that form and remain in one place. Even in this aspect there are similarities, however, for fish tend to congregate in certain areas of the ocean, over topographic rises, in areas of upwelling or where different currents intermix, and in other definable locations. While they are on the hoof, the fish are the property of all; as soon as they are in the fisherman's net or ship, they are his property. The same should logically apply to the miner; however, there are aspects of mining which make it critical that the miner be given some rights to the grounds he is mining. The miner's capital investment is not only in the recovery system, but also in the deposit itself. Fishermen have always been hunters. They spend very little money in an organized effort to find out where the fish are; most good fishing grounds have been known for centuries. Also the fish population is too mobile for a fisherman to search for a school of fish, then stop to evaluate its production potential and develop the best method of catching the fish. He either catches the fish as soon as he finds them or is likely not to find them again. If he does they will probably be in some other fisherman's hold. Not so with sea-floor sediments. They vary in composition and, thus, value from place to place in the ocean. There are some fifteen or so factors or characteristics of a deposit of manganeses nodules which have a bearing on the economics of mining the deposit. These characteristics vary from place to place in the ocean and it is of great importance, both from an economic standpoint and a conservation standpoint, that these characteristics be carefully studied, so that the most efficient mining system practicable can be designed to exploit the deposit. Before any mining operation ensues, therefore, the miner will explore the ocean for several deposits, study those that appear to be capable of yielding the most profit and design his mining system with those nodules in mind. Before he starts to build his mining system, then, the miner will have a substantial capital investment in the deposit or deposits themselves and considerable time will be involved between the initial exploration of a deposit and the mining of that deposit. The miner, thus, would very much like to have some law which grants him the exclusive right to develop and mine a deposit which he has spent substantial amounts of money in exploring. No such law exists at the present time, but, quite possibly, there will be no need for such a law for some time to come even after the first dredge is mining sea-floor sediments. Only a few nations have either the technical capacity to mine the manganese nodules or the industrial system to use the products in any quantity once they are mined and processed. The potential for disputes arising, thus, is considerably reduced.

While it is advantageous for a company to obtain a right to mine the nodules he has explored in a certain area, there is hardly any basis for it to lay claim to ownership with right to sell the deposit in that area. The nodules were discovered by scientists almost 100 years ago and have been considerably explored by other scientists and engineers since that time. Many data have been published describing these deposits. It is fairly well established, that the deposits lie over large areas of the ocean floor and are, in the case of the manganese nodules, quite possibly continuous over thousands of miles of the ocean floor. A mining company could hardly expect to obtain a franchise from any country, nor could a country or agency charge any company a royalty or assess a privilege tax for the mining of any deep-sea deposit. Clearly, the United Nations has no such authority. At the present time, therefore, the nodules or other sediments will probably be considered the property of he who first reduces them to his possession.

There are technical and economic advantages in mining the sediments as close to shore as possible. If the operation is too near shore, a terrific roar will undoubtedly be raised by the nearest nation about robber barons plundering the national patrimony. Such actions are the normal euphemisms employed to cover a request for taxes where they are not clearly due. The mining operator has at least three alternatives: (*1*) move to some other location, (*2*) pay the protection money, or (3) call on his sovereign for protection from interference in his activities on the high seas by a foreign sovereign which is due him by reason of the flag he is flying and under the "Convention on the High Seas".

CHAPMAN (1963) sees this state of affairs as quite satisfactory and sees no reason why a further modification in the law of the sea is required to encourage the beginning of deep-sea mining. For the foreseeable future, capital equipment used in such ventures should be floating and mobile. The law covering the use, responsibilities, and protection of such gear are clearly established in the Convention on the High Seas and in the municipal admiralty law of every maritime state. "The high seas fisheries", Chapman says, "work quite well under these conditions using capital units of comparable size, complexity and value." This is fortunate, for Chapman feels to get a change in existing international law to cover the mining of deep-sea mineral deposits would, in his experience, take a minimum of 10 years, and more likely, 50 years, as well as absorb most of any profits anticipated from the operation.

Chapman concluded his address by saying, "So we fishery people welcome the miners aboard and into that 70% of the world where some freedoms and responsibilities remain. We only pray that they just work hard and make money and don't bother about the law. This generation of fish people has had a sufficiency of international conferences on the law of the sea and it's reasonably satisfied to maintain the status quo, whatever that is."

Such is the law of the sea.

APPENDIX I

STATION OF SAMPLE TITLE LIST

(Explanation of abbreviations in tables)

A.[1]	Atlantis
Alb.[1]	Albatross
Acap.[1]	Acapulco
Blake T	Blake Trawl
BM	British Museum sample no.
Cap.[1]	Capricorn
Car.[1]	Carnegie
Chal.[1]	Challenger (1873–76)
Chub.[1]	Chubasco
Cusp[1]	Cusp
DH	Dredge haul
DWBD[1]	Downwind BAIRD dredge haul. Name of cruise-name of ship-sampling device
DWBG[1]	Downwind BAIRD gravity core
DWHD[1]	Downwind HORIZON dredge haul
DWHG[1]	Downwind HORIZON gravity core
DWP[1]	Downwind photographic
Expl.[1]	Explorer
Fan Bd[1]	Fanfare BAIRD dredge haul
Japan	Station off Japan
JEDS[1]	Japan eastern deep survey
JYN[1]	Japanyon
MP[1]	Mid Pacific
Msn[1]	Monsoon
Muir Smt	Muir Seamount
Naga[1]	Naga
NH[1]	North Holiday
PAS	Pelagic area study
S Clem SV	San Clemente Sea Valley
SOB[1]	Southern Borderland
SW[1]	Swedish deep-sea expedition
Tet.[1]	Tethys
Theta[1]	Theta
Trans.[1]	Transfer
UNK	Unknown cruise
UPWD[1]	Upwind dredge haul
US Navy	United States Navy dredge haul

[1] Names of cruises.

V.[1]	Vema (Name of Lamont Geological Observatory ship)
Vit.[1]	Vitiaz
VS[1]	Vermillion Sea
WHOI	Woods Hole Oceanographic Institution
Wig.[1]	Wigwam

[1] Names of cruises.

APPENDIX II

TABLE OF CONVERSION FACTORS

To convert	Into	Multiply by
inches	cm	2.540
cm	inches	0.3937
ft.	m	0.3048
m	ft.	3.281
miles (statute)	km	1.609
km	miles (statute)	0.6214
fathoms	ft.	6.0
ft.	fathoms	0.167
fathoms	m	1.8288
m	fathoms	0.547
gallons (U.S. liq.)	l	3.785
l	gallons (U.S. liq.)	0.2642
g	lb. (Avoirdupois)	$2.205 \cdot 10^{-3}$
lb. (Avoirdupois)	g	453.59
square miles	km^2	2.590
km^2	square miles	0.3861
g/cm^2	lb./square ft.	2.0481
g/cm^2	metric tons/km^2	10^4
lb./square ft.	short tons/square mile	$1.394 \cdot 10^4$
metric tons/km^2	short tons/square mile	2.855
short tons/square mile	metric tons/km^2	0.351

REFERENCES

AGASSIZ, A., 1901. "Albatross" Expedition preliminary report. *Mem. Museum Comp. Zool., Harvard College, Cambridge*, 26: 1–111.

AGASSIZ, A., 1906. "Albatross" Expedition reports. *Mem. Museum Comp. Zool., Harvard College, Cambridge*, 33: 1–50.

ALTSCHULER, Z. S., CLARKE, R. S. AND YOUNG, E. J., 1958. Geochemistry of uranium in apatite and phosphorite. *U.S. Geol. Surv., Profess. Papers*, 314-D: 90 pp.

AMES, L. L., 1959. The genesis of carbonate apatites. *Econ. Geol.*, 54: 829–841.

ANONYMOUS, 1962. Grab dredge will mine sand-iron. *Eng. Mining J.*, 163: 97.

ANONYMOUS, 1963a. Jet smelting of iron ore. *Iron Age*, 192: 62–63.

ANONYMOUS, 1963b. New giant drills off California. *Oil Gas J.*, 61 (28): 86–87.

ARMSTRONG, E. F. and MIALL, L. M., 1946. *Raw Materials from the Sea*. Chemical Publishing Co., Brooklyn, N. Y., 196 pp.

ARRHENIUS, G., 1959. Sedimentation on the ocean floor. In: P. ABELSON (Editor), *Researches in Geochemistry*, Wiley, New York, N. Y., pp. 1–24.

ARRHENIUS, G., 1963. Pelagic sediments. In: M. N. HILL, E. D. GOLDBERG, C. O' D. ISELIN and W. H. MUNK (Editors), *The Sea, Ideas and Observations on Progress in the Study of the Seas*, Wiley, New York, N. Y., 3: 655–727.

ARRHENIUS, G., BRAMLETTE, M. W. and PICCIOTTO, E., 1957. Localization of radioactive and stable heavy nuclides in ocean sediments. *Nature*, 180: 85–86.

ASCHAN, O., 1932. Über Wasserhumus und seine Beteiligung an der Erzbildung in den nordischen Süssgewässern. *Nachr. Akad. Wiss. Goettingen, Math. Physik. Kl.*, 4 (29): 505–521.

ATWATER, G. I., 1956. Future of Louisiana offshore oil province. *Bull. Am. Assoc. Petrol. Geologists*, 40: 2624–2634.

AUBEL, V. W., 1920. Titaniferous iron sands of New Zealand. *Am. Inst. Mining Met. Engrs., Trans.*, 63: 266–288.

BARDT, H., 1927. Precious metals from seawater. *Brit. Pat.*, 294,655.

BASCOM, W., 1962. The mohole project. In: J. G. STRYKOWSKI (Editor), *Underwater Yearbook 1962*. Underwater Soc. Am., Chicago, Ill., pp. 15–19.

BAUDIN, E. M. L., 1916. Separating gold and silver products from seawater. *French Pat.*, 481, 491.

BAUER, E. and NAGEL, O., 1912. Process for recovering precious metals from very dilute solutions, especially seawater. *Ger. Pat.*, 272,654.

BAUM, A. W., 1960. Fabulous mine in the sea. *Saturday Evening Post*, 1960 (April 23), 4 pp.

BEASLEY, A. W., 1948. Heavy mineral beach sands of southern Queensland. *Proc. Roy. Soc. Queensland*, 59: 109–140.

BECKMANN, W. C., 1960. Geophysical surveying for a channel tunnel. *New Scientist (U.K.)*, 7: 710–712.

BECKMANN, W. C., ROBERTS, A. C. and THOMPSON, K. C., 1962. How underwater seismics aided Thailand tin exploration. *Eng. Mining J.*, 163: 244–247.

BLANCHARD, F. M. and ROMANOWITZ, C. M., 1956. Chain-bucket or bucket-ladder dredges. In: R. PEELE (Editor), *Mining Engineers' Handbook*. Wiley, New York, N. Y., pp. 10-577–10-600.

BLASKETT, K. S. and DUNKIN, H. H., 1948. The occurrence of chromium in ilmenite from Norries Head, New South Wales, and Stradbroke Island, Queensland, Australia. *Univ. Melbourne, Council Sci. Ind. Res., Ore Dressing Investigation*, 337: 6 pp.

BOOS, M. F., 1940. Black beach sands of Guatemala, Central America *Bull. Geol. Soc. Am.*, 51: 1921.

BOUSSINGAULT, F., 1882. Sur l'apparition du manganèse à la surface des roches. *Ann. Chim. et Phys.*, 27: 289–311.

BRAMLETTE, M. N., 1961. Pelagic sediments. In: M. SEARS (Editor), *Oceanography — Publ. Am. Assoc. Advan. Sci.*, 67: 345–366.

BRAMLETTE, M. N. and BRADLEY, W., 1942. Geology and biology of North Atlantic deep-sea cores between Newfoundland and Ireland. *U. S. Geol. Surv., Profess. Papers*, 196: 1–34.

BRUUN, A. F. and WOLFF, T., 1961. Abyssal benthic organisms: Nature, origin, distribution, and influence on sedimentation. In: M. SEARS (Editor), *Oceanography — Publ. Am. Assoc. Advan. Sci.*, 67: 391–397.

BUCHANAN, J. Y., 1890. On the occurrence of sulphur in marine muds and nodules, and its bearing on their mode of formation. *Proc. Roy. Soc. Edinburgh*, 18: 17–39.

BUSER, W., 1959. The nature of the iron and manganese compounds in manganese nodules. In: M. SEARS (Editor), *Intern. Oceanog. Congr. Preprints — Publ. Am. Assoc. Advan. Sci.*, pp. 962–963.

BUSER, W. und GRÜTTER, A., 1956. Über die Natur der Manganknollen. *Schweiz. Mineral. Petrogr. Mitt.*, 36: 49–62.

BUTKEVITSCH, W. S., 1928. The formation of marine iron and manganese deposits and the role of micro-organisms in the latter. *Ber. Wiss. Merresinat., Moscow*, 3: 67–80.

CARLSON, O. J., 1944. Exploitation of minerals in beach sands on the south coast of Queensland. *Queensland Govt. Mining J.*, 49: 223–245.

CARO, J. H. and HILL, W. L., 1958. Characteristics and fertilizer value of phosphate rocks from different fields. *Agr. Food Chem.*, 4: 684–687.

CASPARI, W. A., 1910. The composition and character of oceanic clay. *Proc. Roy. Soc. Edinburgh*, 30: 183–201.

CERNIK, B., 1926. Extracting gold from sea water. *Ger. Pat.*, 490,207.

CERNIK, B., 1927. Extracting gold from sea water. *French Pat.*, 633,998.

CHAPMAN, W. M., 1963. Legal problems in harvesting minerals of the deep sea bed. *Symposia on Economic Importance of Chemicals from the Sea*. Am. Chem. Soc., Div. Chem. Marketing Econ., Washington, D. C., pp. 177–186.

CHEN, P., 1963. Heavy mineral deposits of western Taiwan. *Bull. Geol. Surv. Taiwan*, 4: 13–21.

CHESTERMAN, C. W., 1952. Descriptive petrography of rocks dredged off the coast of central California. *Proc. Calif. Acad. Sci.* 27: 359–374.

CHOW, T. J. and PATTERSON, C. C., 1959. Lead isotopes in manganese nodules. *Geochim. Cosmochim. Acta*, 17: 21–31.

CHOW, T. J. and PATTERSON, C. C., 1962. The occurrence and significance of lead isotopes in pelagic sediments. *Geochim. Cosmochim. Acta*, 26: 263–308.

CLARKE, F. W., 1924. The data of geochemistry. *U. S. Geol. Surv., Bull.*, 770: 832.

COETZEE, C. B., 1957. Ilmenite-bearing sands along the west coast in the Vanrhynsdrop District. *Union S. Africa, Dept. Mines, Geol. Surv. Div. Bull.*, 25: 1–17.

COLLINS, S. V. and KEEBLE, P., 1962. Diamonds from the sea bed. In: J. G. STRYKOWSKI (Editor), *Underwater Yearbook 1962*. Underwater Soc. Am., Chicago, Ill., pp. 12–14.

COOK, G. B. and DUNCAN, J. F., 1952. *Modern Radiochemical Practice*. Oxford Univ. Press, London, 350 pp.

COOPER, L. H. N., 1948. The distribution of iron in water of the western English Channel. *J. Marine Biol. Assoc. U. K.*, 25: 279–313.

CORRENS, W., 1939. Pelagic sediments of the North Atlantic Ocean. In: P. D. TRASK (Editor), *Recent Marine Sediments*. Am. Assoc. Petrol. Geologists, Tulsa, pp. 373–395.

CORRENS, W., 1941. Beiträge zur Geochemie des Eisens und Mangans. *Nachr. Akad. Wiss. Goettingen, Math. Physik. Kl.*, 5: 219.

CRUICKSHANK, M. J., 1962. *The Exploration and Exploitation of Offshore Mineral Deposits*. M. S. Thesis, Colorado School of Mines, Golden, Colo., 185 pp.

CRUICKSHANK, M. J., 1963. *Mining Offshore Alluvials*. Dept. Min. Tech., Univ. Calif., Berkeley, Calif., unpubl. rept.

DIETZ, R. S., 1955. Manganese deposits on the northeast Pacific sea floor. *Calif. J. Mines Geol.*, 51: 209–220.

DIETZ, R. S., EMERY, K. O. and SHEPARD, F. P., 1942. Phosphorite deposits on the sea floor off southern California. *Bull. Geol. Soc. Am.*, 53: 815–848.

DIEULAFAIT, L., 1883. Le manganèse dans les eaux de mers actuelles et dans certain de leur dépôts. *Compt. Rend.*, 96: 718.

DORFF, F., 1935. *Biologie des Eisens und Mangankreislaufes*, Springer, Berlin.

DULIEUX, P. E., 1912. The magnetic sands of the north shore of the Gulf of St. Lawrence. In: *Mining Operations in the Province of Quebec during the Year 1911*. Quebec Dept. Colonization, Mines, Fisheries, pp. 135–156.

EDGERTON, H. E., 1955. Photographing the sea's dark underworld. *Natl. Geograph. Mag.*, 107: 523–537.

EHRLICH, H. L., 1963. Bacteriology of manganese nodules. *Appl. Microbiol.*, 28: 15–19.

ELMENDORF, C. H. and HEEZEN, B. C., 1957. Oceanographic information for engineering submarine cable systems. *Bell System Tech. J.*, 36: 1047–1093.

EL WAKEEL, S. K. and RILEY, J. P., 1961. Chemical and mineralogical studies of deep-sea sediments. *Geochim. Cosmochim. Acta*, 25: 110–146.

EMERY, K. O., 1960. *The Sea Off Southern California*. Wiley, New York, N.Y., 366 pp.

EMERY, K. O. AND DIETZ, R. S., 1950. Submarine phosphorite deposits off California and Mexico. *Calif. J. Mines Geol.*, 46: 7–15.

EMERY, K. O. AND SHEPARD, F. O., 1945. Lithology of the sea floor off southern California. *Bull. Geol. Soc. Am.*, 56: 431–478.

ERICSON, D. B., EWING, M., WOLLIN, G. and HEEZEN, B. C., 1961. Atlantic deep-sea sediment cores. *Bull. Geol. Soc. Am.*, 72: 193–286.

EWING, M., VINE, A. and WORZEL, J. L., 1946. Photography of the ocean bottom. *J. Opt. Soc. Am.*, 36: 307–321.

FLANGAS, W. G. and SHAFFER, L. E., 1960. An application of nuclear explosives to block caving in mining. *Univ. Calif. (Berkeley), Radiation Lab. Rept.*, 5949, T 10-4500: 23 pp.

GARRELS, R. M., 1960. *Mineral Equilibria*. Harper, New York, N. Y., 254 pp.

GIBSON, T. M., 1911. Paystreaks at Nome. *Mining Sci. Press*, 102: 424–427.

GILLSON, J. W., 1951. Deposits of heavy minerals on the Brazilian coast. *Trans. A.I.M.E.*, 187: 685–693.

GOLDBERG, E. D., 1952. Iron assimilation by marine diatoms. *Biol. Bull.*, 102: 243–248.

GOLDBERG, E. D., 1954. Marine geochemistry. 1. Chemical scavengers of the sea. *J. Geol.*, 62: 249–265.

GOLDBERG, E. D., 1957. Biogeochemistry of trace metals. *Geol. Soc. Am., Mem.*, 1: 345–358.

GOLDBERG, E. D., 1960. Phosphatized wood from the Pacific sea floor. *Bull. Geol. Soc. Am.*, 71: 631–632.

GOLDBERG, E. D., 1961a. Chemical and mineralogical aspects of deep-sea sediments. In: L. H. AHRENS (Editor), *Physics and Chemistry of the Earth*. Pitman, Bath, pp. 281–302.

GOLDBERG, E. D., 1961b. Chemistry in the oceans. In: M. SEARS (Editor), *Oceanography — Publ. Am. Assoc. Advan. Sci.*, 67: 583–597.

GOLDBERG, E. D., 1963a. The oceans as a chemical system. In: M. N. HILL, E. D. GOLDBERG, C. O' D. ISELIN and W. H. MUNK (Editors), *The Sea, Ideas and Observations on Progress in the Study of the Seas*. Wiley, New York, N. Y., 2: 3–25.

GOLDBERG, E. D., 1963b. Mineralogy and chemistry of marine sedimentation. In: F. P. SHEPARD, *Submarine Geology*, Harper, New York, N. Y., pp. 436–466.

GOLDBERG, E. D. and ARRHENIUS, G., 1958. Chemistry of Pacific pelagic sediments. *Geochim. Cosmochim. Acta*, 13: 153–212.

GOLDBERG, E. D. and KOIDE, M., 1958. Ionium–thorium chronology in deep-sea sediments. *Science*, 123: 1003.

GOLDBERG, E. D. and KOIDE, M., 1962. Geochronological studies of deep-sea sediments by the ionium–thorium method. *Geochim. Cosmochim. Acta*, 26: 417–450.

GOLDBERG, E. D. and PICCIOTTO, E., 1955. Thorium determinations in manganese nodules. *Science*, 121: 613–614.

GOLDBERG, E. D., MCBLAIR, W. and TAYLOR, K. M., 1951. The uptake of vanadium by tunicates. *Biol. Bull.*, 101: 84.

GOLDSCHMIDT, V. M., 1954. *Geochemistry*. Oxford Univ. Press, London, 730 pp.

GRAHAM, J. W., 1959. Metabolically induced precipitation of trace elements from sea water. *Science*, 129: 1428–1429.

GRAHAM, J. W. and COOPER, S. C., 1959. Biological origin of manganese rich deposits of the sea floor. *Nature*, 183: 1050–1051.

GRIGGS, A. B., 1945. Chromite bearing sands of the southern part of the coast of Oregon. *U.S. Geol. Surv., Bull.*, 945-E: 113–150.

GRUNER, J. W., 1922. The origin of sedimentary iron formations: the Biwabik Formation of the Mesabi Range. *Econ. Geol.*, 17: 407–466.

GRÜTTER, A. und BUSER, W., 1957. Untersuchungen an Mangansedimenten. *Chimia (Aarau)*, 11: 132–133.

HABER, F., 1927. Das Gold im Meerwassers. *Z. Angew. Chem.*, 40: 303–317.

HAMILTON, E. L., 1956. Sunken islands of the Mid-Pacific Mountains. *Geol. Soc. Am., Mem.*, 64: 97 pp.

HAMMOND, R. P., 1962. Large reactors may distill sea water economically. *Nucleonics*, 20: 45–50.

HANNA, G. D., 1952. Geology of the continental slope off central California. *Proc. Calif. Acad. Sci.*, 27: 325–374.

HANSON, A. W., 1958. Mining soluble minerals by circulation of a solvent. *U. S. Pat.*, 2, 850, 270.

HARVEY, E. E., 1950. *Spectrochemical Procedures*. Appl. Res. Labs., Glendale, Calif., 402 pp.

HARVEY, H. W., 1960. *The Chemistry and Fertility of Sea Waters*. Cambridge Univ. Press, London, 240 pp.

HEEZEN, B. C., THARP, M. and EWING, M., 1959. The floor of the oceans. 1. The North Atlantic. *Geol. Soc. Am., Spec. Papers*, 65: 122 pp.

HESS, H. H., 1946. Drowned ancient islands of the Pacific basin. *Am. J. Sci.*, 244: 772–791.

HIGAZY, R. A. and NAGUIB, A. G., 1958. Egyptian monazite-bearing black sands. *Proc. U.N. Intern. Conf. Peaceful Uses At. Energy, 2nd, Geneva, 1958*, 2: 658–662.

HOPKINS, D. M., MACNEIL, F. S. and LEOPOLD, E. B., 1960. The coastal plain at Nome, Alaska. *Intern. Geol. Congr., 21st, Copenhagen, 1960, Rept. Session, Norden.*

HORTIG, F. J., 1958. California offshore oil, present and future. *Ann. Meeting Am. Assoc. Petrol. Geologists, Pacific Section, 35th, Los Angeles, Calif., 1958* (unpubl. talk).

JONES, E. J., 1887. On some nodular stones obtained by trawling off Colombo in 675 fathoms of water. *J. Asiatic Soc. Bengal*, 56: 209–212.

JOHNSON, D. W., 1919. *Shore Processes and Shoreline Development*. Wiley, New York, N. Y., 584 pp.

KALINENKO, V. O., 1949. The origin of Fe–Mn concretions. *Mikrobiologiya*, 18: 528–532.

KRAUSKOPF, K. B., 1956. Factors controlling the concentrations of thirteen rare metals in sea water. *Geochim. Cosmochim. Acta*, 9: 1–32.

KRAUSKOPF, K. B., 1957. Separation of manganese from iron in sedimentary processes. *Geochim. Cosmochim. Acta*, 12: 61–84.

KRUMBEIN, W. C. and GARRELS, R. M., 1952. Origin and classification of chemical sediments in terms of pH and oxidation reduction potentials. *J. Geol.*, 60: 1–33.

KUENEN, PH. H., 1950. *Marine Geology*. Wiley, New York, N. Y., 568 pp.

LEE, C. O., BARTLETT, Z. W. and FEIERABEND, R. H., 1960. The Grand Isle mine. *Mining Engr.*, 120: 587–590.

LJUNGGREN, P., 1953. Some data concerning the formation of manganiferous and ferriferous bog ores. *Geol. Fören. Stockholm Förh.*, 75: 277–297.

REFERENCES

LYDON, P. A., 1957. Titanium. In: *Mineral Commodities of California — Calif. Dept. Nat. Resources, Div. Mines, Bull.*, 176: 647–654.

MADDREN, A. G., 1919. The beach placers of the west coast of Kodiak Island, Alaska. In: *Mineral Resources of Alaska in 1917 — U. S. Geol. Surv., Bull.*, 692: 299–319.

MANNING, P. D. V., 1936. Oceans of raw material for magnesium compounds. *Chem. Met. Eng.*, 43: 116–120.

MANNING, P. D. V., 1938. Magnesium metal and compounds from sea water and bitterns. *Chem. Met. Eng.*, 45: 478–482.

MANSFIELD, G. R., 1940. The role of fluorine in phosphate deposition. *Am. J. Sci.*, 238: 863–879.

MARIES, A. C. and BECKMANN, W. C., 1961. A new geophysical method for the exploration of undersea coal fields. *Mining Engr.*, 120: 262–276.

MARTENS, J. H. C., 1928. Beach deposits of ilmenite, zircon, and rutile in Florida. *Florida Geol. Surv. Ann. Rept*, 19: 124–154.

MCILHENNY, W. F. and BALLARD, D. A., 1963. The sea as a source of dissolved chemicals. *Symposia on Economic Importance of Chemicals from the Sea*. Am. Chem. Soc., Div. Chem. Marketing Econ., Washington, D. C., pp. 122–131.

MCKELVEY, V. E and BALSLEY, J. R., 1948. Distribution of coastal black sands in North Carolina, South Carolina and Georgia, as mapped from an airplane. *Econ. Geol.*, 43: 518–524.

MENARD, H. W., 1959. Geology of the Pacific sea floor. *Experientia*, 15: 205–213.

MENARD, H. W., 1960. Consolidated slabs on the floor of the eastern Pacific. *Deep-Sea Res.*, 7: 35–41.

MENARD, H. W. and SHIPEK, C. J., 1958. Surface concentrations of manganese nodules. *Nature*, 182: 1156–1158.

MERO, J. L., 1952. Manganese. *N. Dakota Engr.*, 27: 28–32.

MERO, J. L., 1959. *The Mining and Processing of Deep-Sea Manganese Nodules*. Univ. Calif. (Berkeley) Inst. Marine Res., 96 pp.

MERO, J. L., 1960a. *The Economics of Mining Phosphorite from the California Borderland Area*. Univ. Calif. (Berkeley) Inst. Marine Res., 150 pp.

MERO, J. L., 1960b. Minerals on the ocean floor. *Sci. Am.*, 203: 64–72.

MERO, J. L., 1960c. Mineral resources on the ocean floor. *Mining Congr. J.*, 46: 48–53.

MERO, J. L., 1962. Ocean-floor manganese nodules. *Econ. Geol.*, 57: 747–767.

MERTIE, J. B., 1939. Platinum deposits of the Goodnews Bay District, Alaska. *U. S. Geol. Surv., Bull.*, 910-B: 30 pp.

MERTIE, J. B., 1940. The Goodnews platinum deposits, Alaska. *U. S. Geol. Surv., Bull.*, 918: 97 pp.

MURRAY, J. and IRVINE, R., 1894. On the manganese oxides and manganese nodules in marine deposits. *Trans. Roy. Soc. Edinburgh*, 37: 712–742.

MURRAY, J. and LEE, G. V., 1909. The depth and marine deposits of the Pacific. *Mem. Museum Comp. Zool., Harvard College, Cambridge*, 38: 1–171.

MURRAY, J. and RENARD, A., 1891. Report on deep-sea deposits. In: C. WYVILLE THOMSON (Editor), *Report on the Scientific Results of the Voyage of H.M.S. "Challenger"*. Eyre and Spottiswoode, London, 5: 1–525.

NICCALI, E., 1925. Potassium salts from sea water. *Brit. Pat.*, 247,405.

NIINO, H., 1959. Manganese nodules from shallow water off Japan. In: M. SEARS

(Editor), *Intern. Oceanog. Congr. Preprints — Publ. Am. Assoc. Advan. Sci.*, pp. 646–647.

O'BRIEN, M. P. and HICKOX, G. H., 1937. *Applied Fluid Mechanics.* McGraw-Hill, New York, N. Y., 467 pp.

OFFICER, C. B., 1959. Continuous seismic profiler aids marine exploration. *World Oil*, 148: 107–110.

PALMER, P., 1960. Sulphur under the sea. *Sea Frontiers*, 6: 210–217.

PARTRIDGE, F. C., 1938. Note on the Durban beach sands. *Trans. Geol. Soc. S. Africa*, 41: 175.

PEPPER, J. F., 1958. Potential mineral resources of the continental shelves of the western hemisphere. In: *An Introduction to the Geology and Mineral Resources of the Continental Shelves of the Americas — U. S. Geol. Surv., Bull.*, 1067: 43–65.

PETTERSON, S. O., 1928. Production of gypsum and magnesium from sea water. *Swed. Pat.*, 65,434.

PETTERSSON, H., 1943. Manganese nodules and the chronology of the ocean floor. *Medd. Oceanog. Inst. Göteborg, Ser. B*, 2: 1–39.

PETTERSSON, H., 1945. Iron and manganese on the ocean floor. *Medd. Oceanog. Inst. Göteborg, Ser. B*, 3: 1–37.

PHELPS, W. B., 1940. Heavy minerals in the beach sands of Florida. *Proc. Florida Acad. Sci.*, 5: 168–171.

POLDERVAART, A., 1955. Chemistry of the earth's crust. In: A. POLDERVAART (Editor), *The Crust of the Earth — Geol. Soc. Am., Spec. Papers*, 62: 119–144.

PRATT, W. E., 1951. *Fuel Reserves of the United States.* Statement prepared at request of Comm. on Interior and Insular Affairs, U. S. Senate.

PRATT, W., 1961. The origin and distribution of glauconite and related clay mineral aggregates off southern California. *Proc. Natl. Coastal Shallow Water Res. Conf., 1st.*, 1961, pp. 656–658.

RANKAMA, I. and SAHAMA, T. G., 1950. *Geochemistry.* Univ. of Chicago Press. Chicago, Ill., 912 pp.

RAO, C. B., 1957. Beach erosion and concentration of heavy minerals sands. *J. Sediment. Petrol.*, 27: 143–147.

REVELLE, R. R., 1944. Marine bottom samples collected in the Pacific Ocean by the "Carnegie" on its seventh cruise. *Carnegie Inst., Wash. Publ.*, 556: 1–182.

REVELLE, R. AND EMERY, K. O., 1951. Barite concretions from the ocean floor. *Bull. Geol. Soc. Am.*, 62: 707–724.

REVELLE, R., BRAMLETTE, M., ARRHENIUS, G. and GOLDBERG, E. D., 1955. Pelagic sediments of the Pacific. In: A. POLDERVAART (Editor), *Crust of the Earth — Geol. Soc Am., Spec. Papers*, 62: 221–236.

RICHARDS, A. F., 1958. Transpacific distribution of floating pumice from Isla San Benedicto, Mexico, *Deep-Sea Res.*, 5: 29–35.

RILEY, J. P. and SINHASENI, P., 1958. Chemical composition of three manganese nodules from the Pacific Ocean. *J. Marine Res. (Sears Found. Marine Res.)*, 17: 466–482.

RITTER, J. H., 1938. Seawater filter utilizing mercury for recovering gold, etc. *U. S. Pat.*, 2,097,645.

ROMANOWITZ, C. M., 1962. The dredge of tomorrow. *Eng. Mining J.*, 163: 84–91.

SEATON. M. Y., 1931. Bromine and magnesium compounds drawn from western bays and hills. *Chem. Met. Eng.*, 38: 638–641.

SHEPARD, F. P., 1941. Non-depositional environments off the California coast. *Bull. Geol. Soc. Am.*, 52: 1869–1886.

SHEPARD, F. P., 1959. *The Earth Beneath the Sea*. The Johns Hopkins Univ. Press, Baltimore, Md., 275 pp.

SHEPARD, F. P., 1963. *Submarine Geology*. Harper, New York, N. Y., 557 pp.

SHIGLEY, C. M., 1951. Minerals from the sea. *J. Metals*, 3: 25–29.

SHIPEK, C., 1960. Photographic study of some deep-sea floor environments in the eastern Pacific. *Bull. Geol. Soc. Am.*, 71: 1067–1074.

SKORNYAKOVA, N. S., 1960. Manganese concretions in sediments of the northeastern Pacific Ocean. *Dokl. Akad. Nauk S.S.S.R.*, 130: 653–656.

SKORNYAKOVA, N. S. and ZENKEVITCH, N., 1961. The distribution of iron–manganese nodules in surficial sediments of the Pacific Ocean. *Okeanologiya*, 1: 86–94.

SKORNYAKOVA, N. S., ANDRUSCHENKO, P. F. and FOMINA, L. S., 1962. The chemical composition of iron–manganese nodules of the Pacific Ocean. *Okeanologiya*, 2: 264–277.

SMIRNOV, A. I., 1957. The problem of the genesis of phosphorite. *Dokl. Akad. Nauk S.S.S.R.*, 119: 53–56.

SOKOLOVA, M. N., 1959. On the distribution of deep-water bottom animals in relation to their feeding habits and the character of sedimentation. *Deep-Sea Res.*, 6: 1–4.

SPÄRCK, B., 1956. The density of animals on the ocean floor. In: A. BRUUN (Editor), *The "Galathea" Deep-Sea Expedition*. Allen and Unwin, London, pp. 196–201.

STOCES, B., 1925. Extracting gold from sea water with antimony sulfide or other metallic sulfides. *Brit. Pat.*, 273,346.

SVERDRUP, H. U., JOHNSON, M. W. and FLEMING, R. H., 1942. *The Oceans, Their Physics, Chemistry and General Biology*. Prentice-Hall, Englewood Cliffs, N. J., 1087 pp.

SWAIN, P., 1963. Offshore drilling now circles the globe. *Oil Gas J.*, 61: 130–134.

TAGGART, A. F., 1945. *Handbook of Mineral Dressing*. Wiley, New York, N. Y., 1905 pp.

TATSUMOTO, M. and GOLDBERG, E. D., 1959. Some aspects of the marine geochemistry of uranium. *Geochim. Cosmochim. Acta*, 17: 201–208.

TERRY, R. D., 1964. *Oceanography, Its Tools, Methods, Resources and Applications*. MacMillan, New York, N. Y., in press.

THOMASSON, E. M., 1958. Problems of petroleum development on the continental shelf of the Gulf of Mexico. In: *An Introduction to the Geology and Mineral Resources of the Continental Shelves of the Americas — U. S. Geol. Surv., Bull.*, 1067: 67–92.

TIPPER, G. H., 1914. The monazite sands of Travancore. *Records Geol. Surv., India*, 44: 186–196.

TROXEL, B. W., 1957. Abrasives. *Mineral Commodities of California — Calif. Dept. Nat. Resources Div. Mines, Bull.*, 176: 23–28.

TRUMBULL, J., 1958. Continents and ocean basins and their relation to continental shelves and continental slopes. In: *An Introduction to the Geology and Mineral Resources of the Continental Shelves of the Americas — U. S. Geol. Surv., Bull.*, 1067: 1–26.

TWENHOFEL, W. H., 1943. Origin of the black sands of the coast of southwest Oregon. *Oregon Dept. Geol. Mineral Ind., Bull.*, 24: 25 pp.
U. S. BUREAU OF MINES STAFF, 1956. *Mineral Facts and Problems — U. S. Bur. Mines, Bull.*, 556: 1042 pp.
VERPLANCK, W. E., 1958. Salt in California. *Calif. Dept. Nat. Resourses, Div. Mines, Bull.*, 175: 168 pp.
VIENNE, G., 1949. Extracting iodine from seawater. *French Pat.*, 945,357.
VON BUTTLAR, H. und HOUTERMANS, G., 1950. Photographische Bestimmung der Aktivitätverteilung in einer Manganknolle der Tiefsee. *Naturwissenschaften*, 37: 400–440.
WALFORD, L. A., 1958. *Living Resources of the Sea*. Ronald, New York, N. Y., 321 pp.
WARDANI, E., 1959. Marine geochemistry of germanium and the origin of Pacific pelagic clay minerals. *Geochim. Cosmochim. Acta*, 15: 237–244.
WILLIS, J. P. and AHRENS, L. H., 1962. Some investigations on the composition of manganese nodules, with particular reference to certain trace elements. *Geochim Cosmochim. Acta*, 26: 751–764.
WORZEL, J. L. and SHURBET, G. L., 1955. Gravity interpretations from standard oceanic and continental crustal sections. In: A. POLDERVAART (Editor), *The Crust of the Earth — Geol. Soc. Am., Spec. Papers*, 62: 87–100.
ZENKEVITCH, L. A., 1961. Certain quantitative characteristics of the pelagic and bottom life of the ocean. In: M. SEARS (Editor), *Oceanography — Publ. Am. Assoc. Advan. Sci.*, 67: 323–335.
ZENKEVITCH, N. and SKORNYAKOVA, N. S., 1961. Iron and manganese on the ocean bottom. *Natura (U.S.S.R.)*, 3: 47–50.

INDEX

Abalone shells, 57
Abyssal hills, 144
Act, Outer Continental Shelf Lands, 283
Act, Submerged Lands, 283
Africa, 12, 13, 59, 66, 70, 73, 102, 122, 251, 281
Agassiz, A., 127
Agulhas Bank, 59, 66
Air-lift dredge, 243, 251
Akranes, 17
Alaska, 8, 12, 14–15, 78, 80, 102
"Albatross" Expedition, 140, 141
Algae, 51, 107
Alkali, 43
Alumina, 74, 112–116, 118, 121, 124, 152, 179, 182–221, 222, 238–241, 277
Aluminum, 25, 44, 47, 123, 126, 180, 235, 275, 278
Anatase, 152, 153
Aniline, 32
Animal activity and nodule formation, 144–145
Animal debris, 117
Antimony, 25, 47
Anvil Creek, 8, 78
Apatite, 153
Argentina, 12, 58
Argon, 25
Ariake Bay, 16, 246
Arrhenius, S., 41
Arsenic, 25
Arsenic oxide, 47
Asia, 4, 16, 20, 57, 58, 72, 73, 77, 78, 98, 102, 147, 228, 246, 247, 250
Atlantic, 13, 17, 41, 79, 97, 103, 104, 105, 113, 116, 124, 125, 128, 138, 140, 141, 142, 162, 163, 178, 180, 234, 236, 237, 238–240, 271
Australia, 12, 19, 57, 58, 73, 122, 236

Authigenic minerals, 24

Backshore, 8, 9
Bacteria, 137, 148, 149, 150, 228
Bacterial action, 229
Baja California, 12, 231, 232
Balard, A. J., 31
Bangka, 79
Barge, 254
Barite, 152
Barite concretions, 76
Barium, 25, 51, 76, 124, 180, 182–221, 229, 231, 235, 238–241
Barium sulphate, 46
Barium sulphate concretions, 75–76
Basaltic layer, 84, 86
Bathyscaph, 89, 106, 252
Beach deposits, type of minerals, 7–8
Beach deposits, type of sediment transport, 6–7
Beaches, 1, 6–23, 243–252, 280
Beaches, concentration of heavy minerals, 8–11
Bell Island, 97
Berm, 9
Bermuda, 162
Beryllium, 24, 181
Billiton, 79
Biological activity, 24
Biotic processes, 2
Biotite, 153
Bismuth, 26, 181
Bitterns, 29, 31
Black sand, 7
Black Sea, 77
Blake Plateau, 132, 134, 161, 236
Blue mud, 106, 141
Borate ion, 26
Borax, 46
Borneo, 102

INDEX

Boron, 25, 44, 180, 235
Brazil, 12, 19, 281
Breakers, 9
Brine, 29
Bromide ion, 26
Bromine, 3, 25, 27, 29, 31, 44, 46
Bromine-extraction plant, 32
Bucket, clamshell, 15, 22, 245, 253
Bucket-ladder dredge, 243, 247–249
Bucket line, 16
Bucket-line dredge, 243

Cadmium, 26, 47, 152, 181
Caesium, 25
Calcareous ooze, 107, 111–115, 140, 169–171, 182–221
Calcareous sand, 57
Calcareous shell deposits, 55
Calcium, 2, 25–27, 37, 124, 152, 180, 182–221, 231, 235, 238–241
Calcium carbonate, 7, 17, 112–116, 118, 179, 222
Calcium chloride, 31
Calcium oxide, 65, 66, 74, 121
Calcium phosphate, 60, 74, 112, 115, 118, 179, 236
Calcium sulphate, 27, 29, 31, 37, 46, 112, 113, 115, 179
California, 12, 18, 19, 28, 31, 32, 34–36, 38, 39, 41, 43, 49, 55, 56, 57, 59, 60, 67, 69–73, 75–77, 98, 102, 282, 289
Canada, 250
Cape Coast, 13
Cape of Good Hope, 59, 70
Cape Romano, 56
Cape San Lucas, 231
Carbon, 25, 179
Carbonate, 50
Carbonate fluorapatite, 64
Carbonate ion, 26
Carbonates, silica, 64
Carbon dioxide, 65, 66
Carlsbad, 19
Carolinas, 236
Caspian Sea, 4
Cassiterite, 7
Caustic soda, 13
Cement rock sands, 16–18

Central America, 12, 72
Cerium, 25, 181
Cetacean ear bones, 117–119, 137
Ceylon, 12, 18
Chalcanthite, 7
"Challenger" Expedition, 59, 70, 74, 123, 127, 140, 141, 176, 177
Chemical scavengers, 149–151
Chile, 12, 43, 288
China, 73
Chloride, 26
Chlorine, 25, 26, 27, 32
Chromite, 12, 18
Chromium, 26, 51, 124, 180, 235
Churn drilling, 83
Clamshell bucket, 15, 22, 245, 253
Clamshell dredge, 243, 247
Clay, 135
Clay minerals, 50
Climate, 15
Clipperton Fracture Zone, 157
Coal, 98
Cobalt, 25, 47, 51, 123, 124, 126, 149, 151, 152, 180, 182–221, 222, 223, 226–233, 235, 238–241, 271, 272, 275, 278–280
Coccolith ooze, 107, 114
Coccolithophoridae, 107
Coccoliths, 111
Cocos Island, 115
Collophane, 62, 64
Colombo, 75
Columbite, 7
Constituent, insoluble in hydrochloric acid, 65, 115
Continental resources, 4
Continental shelf, 1, 53–83, 283, 288, 289
Continental shelf, surficial deposits, 55
Continental slope, 53
Convention on the High Seas, 293
Cooling waters, 49
Copper, 2, 25, 44, 47, 50, 51, 120, 123–126, 149, 150, 152, 180, 182–221, 222, 223, 226–233, 235, 238–241, 271, 275, 278, 280
Coral debris, 182–221
Coral sand, 115, 182–221

INDEX 307

Core drilling, 83
Coring, 16, 17, 22
Cosmic spherules, 121, 123
Costa Rica, 12
Courtois, B., 42
Crevices, 9
Crystallizing pond, 29
Curacao Islands, 66
Currents, ocean-floor, 142–144
Currents, tidal, 9, 15
CUSS, 99, 100

Dead Sea, 4
Deep-sea drag dredge, 252, 253–260
Deep-sea floor, 103–241
Deep-sea hydraulic dredge, 252, 260–272
Diabase, 74
Diamond, 7, 8, 12–14, 78, 79, 251
Diatom ooze, 107, 109, 116, 117
Diatoms, 64
Diving, 22
Dolomite, 37, 39
Drag-bucket dredge, 243
Drag dredge, 244, 252–260
Drag dredge, deep-sea, 252–260
Drag dredging, 155
Drag-lines, 243
Drake Passage, 163
Dredge, air-lift, 243, 251
Dredge, bucket-ladder, 243, 247–249
Dredge, bucket-line, 243
Dredge, clamshell, 243, 247
Dredge, drag-bucket, 243
Dredge, hydraulic, 243, 244, 249, 252, 260–272
Dredge, ladder-bucket, 243, 244
Dredge, wire-line, 243, 245–247
Drilling, 22

Echogram, 20, 21, 81
Economics of ocean mining, 273–280
Ecuador, 102, 281, 282, 288
Egypt, 12, 102
Electrostatics, 50
El Segundo, 19
Encinitas, 19
England, 34, 39, 98

English Channel, 41
Evaporation pond, 28

Faxa Bay, 17
Feldspar, 7, 64, 74, 111
Ferric oxide, 74, 112, 113, 115, 116, 118, 121, 150, 151, 179
Ferromagnesian minerals, 64
Ferrous oxide, 74
Finland, 96
Fiords, 77
Fish bones, 50, 51
Fish oil, 13
Fish skeletal remains, 120
Fissures, 9, 19
Florida, 12, 19, 56, 66, 72, 101, 161, 236
Florida straits, 49
Fluorine, 25, 60, 65, 66, 118
Foraminifera, 64, 66, 74, 76, 107, 137, 147
Foreshore, 9
Formosa, 12, 72
Forty Mile Bank, 62–66, 68–70, 154, 289
France, 41, 281
Francolite, 62, 64
Frasch process, 86, 87, 93, 95
Freeport, 32, 34–36, 56
Fuji volcanic zone, 147, 228

Gallium, 26, 180, 235, 278
Galveston Bay, 55, 56
Garnet, 14
Gas, use of, 20
Geomorphology, 11, 22
Germanium, 26, 51, 120, 181
Germany, 58
Glasgow, 43
Glauconite, 64, 73–75
Glauconite deposits, 55
Globigerina ooze, 74, 107, 109, 111–114, 141
Gneiss, 74
Goethite, 152, 153
Gold, 7, 8, 12, 14, 15, 18, 22, 26, 27, 39, 44, 46, 50, 51
Grab bucket, 16, 245

Grand Island, 15
Grand Isle, 15, 91, 93, 95
Granite, 74
Granitic layer, 84
Gravel, 77–78
Gravity cores, 165
Grease belt, 13
Great Britain, 57, 58, 281
Green mud, 106, 141
Guadelupe, 85
Guatemala, 12
Gulf Coast, 55, 56, 92, 96, 98, 101
Gulf of California, 70, 232, 234
Gulf of Finland, 96
Gulf of Mexico, 16, 35, 55, 91, 92, 101, 106, 282
Gulf of Tehuantepec, 61
Gulf Stream, 236
Guyot, 103
Gypsum, 7, 31, 37

Haber, F., 41
Hard rock, 1
Hawaii, 135
Heavy-media-separation, 13
Heavy minerals, 7, 8–11, 12, 18, 19
Helium, 26
Helsinki, 96
Hemipelagic sediments, 109
High seas, 285, 293
Horizon nodule, 135, 137
Hornblende, 74, 123, 153
Hydraulic dredge, 243, 244, 249, 252, 260–272
Hydraulic dredge, deep-sea, 252, 260–272
Hydraulic mining, 16
Hydraulic suction, 17–18
Hydrobromic acid, 34
Hydrochloric acid, 37
Hydrofluoric acid, 46

Ice Ages, 8, 11
Iceland, 17, 28
Idaho, 66
Igneous rocks, 146, 147
Ilmenite, 7, 12, 18, 19
India, 12, 19

Indian Ocean, 58, 75, 103, 105, 114–116, 120, 125, 129, 178, 234, 236, 241
Indium, 25
Indonesia, 76, 78, 79, 248
Iodine, 25, 27, 42, 43, 46, 51, 52
Ion exchange, 48
Ionium, 154
Iron, 25, 50, 51, 97, 121, 124, 126, 147–150, 152, 154, 180, 182–221, 222, 223, 230–233, 235, 238–241, 277, 278
Iron hydroxides, 149
Iron ore, 16, 20, 250
Iron peroxides, 127
Iron sands, 16
Isthmus of Tehuantepec, 101

Japan, 16, 20, 57, 58, 72, 73, 98, 102, 147, 228, 246, 247, 250
Jefferson, Th., 286
Jet corer, 22
Jussaro Island, 96
Kai Islands, 76
Kodiak, 12
Krakatoa eruption, 148
Krypton, 26
Kure Beach, 32
Kuskokwim Bay, 78, 80
Kyushu Island, 16

Ladder-bucket, 19
Ladder-bucket dredge, 243, 244
Lanthanum, 26, 180, 235
Law of the sea, 284, 287–293
Lead, 26, 50, 51, 120, 123, 124, 149, 180, 182–221, 222, 223, 227, 229, 235, 238–241, 275, 277, 278
Legal aspects of ocean mining, 273, 280–293
Libya, 102
Lime, 16
Limonite, 152
Lithiophorite, 152
Lithium, 25, 51
Lithium hydroxide, 46
Long Beach, 77
Longshore currents, 9
Los Angeles County, 71

Louisiana, 16, 93, 95, 101, 282, 283
Lüderitz, 13

Madagascar, 115
Magnesite, 37
Magnesium, 3, 25–27, 29, 34, 44, 46, 56, 125, 180, 235, 277, 278
Magnesium carbonate, 65, 112, 113, 115, 179
Magnesium chloride, 31, 37
Magnesium compounds, 39, 46
Magnesium hydroxide, 37, 39
Magnesium oxide, 74, 116, 118, 121
Magnesium sulphate, 31, 37
Magnetic surveying, 14, 16, 22, 80, 96
Magnetite, 7, 12, 14, 16, 18, 22, 74, 80, 96, 111, 123, 250
Manganese, 2, 25, 44, 50, 51, 123–126, 140, 147–150, 152, 154, 180, 182–221, 222, 223, 226, 231, 233–235, 238–241, 271, 272, 275, 278, 279, 280
Manganese dioxide, 43, 111, 127, 139, 179
Manganese dioxide grains, 141
Manganese nodules, 107, 119, 121, 127–241, 252, 271–272, 277–279
Manganese oxide, 47, 62, 116, 118, 121
Manganese peroxide, 127
Manganite, 135, 152
Manganous sulfide, 148
Mantle of the earth, 84
Marianas Trench, 106
Marine beaches, 1, 6–23, 243–252, 280
Marine organisms, concentration of elements by, 50–52
Marine resources, 3, 4
Mass mortality, 59, 61
Matagorda Bay, 56
Mercury, 26, 181
Mexico, 72
Mica, 7, 74, 123, 153
Michigan, 31
Mid-Atlantic Ridge, 103
Milk of lime, 35
Mineral, 2
Mineral resources, 2
Minerals in sea water, 24–52
Mining, hydraulic, 16

Mining methods, 242–272
Miocene, 60
Mississippi, 106
Moho, 85
Mohole Project, 100
Mohorovičić discontinuity, 85
Molluscs, 107
Molybdenum, 25, 44, 51, 124, 126, 149, 151, 152, 180, 182–221, 235, 238–241, 275, 277, 278
Molybdenum oxide, 46
Monazite, 7, 12, 18–20
Monterey, 12, 18
Montpellier, 31
Morocco, 58, 66
Moss Landing, 38, 39, 41
Mother of pearl shells, 57
Mud, 55
Mud eaters, 144

Natashquan, 12
Neon, 26
Newfoundland, 97
Newport Bay, 28
New South Wales, 12, 19
Newton's Law, 263
New Zealand, 19, 73, 227
Nickel, 25, 44, 47, 51, 120, 121, 123–126, 149, 151, 152, 180, 181, 182–221, 222, 223, 226–233, 235, 238–241, 271, 272, 275, 278, 279, 280
Nigeria, 102
Niobium, 26, 181
Nitrogen, 25
Nodule formation, animal activity, 144–145
Nome, 8, 12, 14–15, 78
Nontronite, 152
North America, 8, 12, 14–15, 16, 18, 19, 28, 31, 32, 34–36, 37–39, 41, 43, 49, 55–67, 69–73, 75, 78, 80, 98, 101, 102, 161, 225, 227, 231–233, 236, 250, 274, 282, 289
North Carolina, 12, 32, 42
North Sea, 102, 281
Norton Sound, 15
Norway, 77
Nova Scotia, 98

Nuclear method, 86

Ocean-bottom currents, 68, 142–144
Ocean-currents, 271
Ocean-floor currents, 68, 142–144
Offshore, 9
Offshore placers, 243, 252
Ohio, 31
Olifante River, 13
Oolites, 63
Oozes, 107
Opal, 152
Open-pit methods, 243
Orange peel, 245
Orange River, 8, 78, 79
Oregon, 12, 18, 72, 102
Organic constituent, 65, 66
Organic sediments, 76–77
Outer Banks, 65
Outer Continental Shelf Lands Act, 283
Oxidizing atmosphere, 143, 154
Oxygen, 179
Oyster shells, 16, 55

Pacific, 16, 57, 58, 72, 74, 102, 103, 105, 114–116, 120, 123–125, 127, 129–132, 140, 157, 163, 164, 169–171, 173–175, 178–180, 182–221, 222, 223, 226, 228, 229, 232, 234, 271, 276, 278, 279
Palagonite, 121, 123
Patton Escarpment, 65
Pearl oysters, 57
Pelagic sediments, 106–110
Persian Gulf, 102
Peru, 102, 288
Petroleum, 4, 98–102, 252, 282
Phillipines, 72, 73, 134
Phillipsite, 120, 121, 137
Phosphate, 47, 135, 275
Phosphoria, 64
Phosphoria sediments, 60
Phosphorite, 12, 57, 165
Phosphorite deposits, 55
Phosphorite nodules, 57–73, 252
Phosphorus, 16, 25, 57, 182–221, 238–241
Phosphorus pentoxide, 58, 65, 66, 71, 116, 118

Photography, 155–165, 166, 167
Pipe line, 13
Piston coring, 83
Placer deposits, 78–80
Plagioclase, 123
Platinum, 7, 18, 24, 78, 80
Pleistocene, 14
Pliocene, 14, 76
Point Reyes, 70
Portland cement, 16
Portugal, 73
Potash, 31, 43, 91, 121, 182–221, 235, 238–241
Potassium, 25–27, 43, 44, 52, 124, 180
Potassium chloride, 46
Potassium feldspars, 153
Potassium oxide, 74, 116, 121
Potholes, 19
Protactinium, 26
Psilomelane, 152
Pteropod ooze, 107, 109, 114, 115, 141
Pteropods, 111
Pumice, 111, 129, 133, 157, 236
Pyrite, 77
Pyroxene, 153

Quartz, 7, 12, 14, 64, 74, 111, 123, 153
Quebec, 12
Queensland, 12, 19

Radiolarian ooze, 107, 109, 115, 116, 141, 157, 169–171, 182–221
Radiolarians, 64
Radium, 26, 153, 154, 181
Radon, 26
Rare earths, 151, 277
Red clay, 107, 109, 114, 120, 121, 122–127, 141, 159, 178, 182–221, 232
Red mud, 141
Redondo, 12
Redondo Beach, 18
Redondo Canyon, 65
Resources, continental, 4
Resources, marine, 3, 4
Resources, mineral, 2
Reynold's number, 263
Rhabdoliths, 14
Ripple marks, 143

INDEX 311

River valleys, drowned, 78–83
Robots, 88, 89
Rotary drilling, 83
Rubidium, 25, 51
Russia, 30
Rutile, 12, 18, 19, 152, 153

Salt, 3, 26, 27
Salt dome, 91, 252
Salmon River, 78, 80
San Clemente Island, 76
Sand, 55, 77–78
Sand bar, 9
San Diego, 62, 64, 70, 154, 289
San Diego Bay, 28
San Francisco, 70, 77
San Francisco Bay, 16, 28, 41, 55
San Pedro Bay, 78
Santa Barbara Basin, 76
Santa Monica, 70
Santa Monica Canyon, 65
Scandium, 26, 124, 180, 235
Scavengers, chemical, 149–151
Scheelite, 7
Scotland, 73
Scour marks, 142, 162
Sea and life on earth, 1
Sea as a source of minerals, 1
Sea level, changes of, 8, 9, 11
Seamounts, 103, 230
Sea water, 1
Sea-water conversion, 43–45
Sea water, elements, 25–26
Sea water, suspended matter in, 50
Seaweeds, 27, 42, 43, 52
Sediment waves, 162
Seismic surveying, 11, 14, 92
Seismology, 85
Selenium, 25, 46
Senegal, 12
Shark teeth, 117, 118, 119, 137
Shelf, continental, 1, 53–83, 89–96, 283, 288, 289
Shelf sediments, 109
Shells of a nodule, 231
Sherritt–Gordon process, 271
Silica, 50, 64, 135, 148, 152, 182–221, 222, 231, 238–241

Silica sand, 18
Silicates, 7
Siliceous ooze, 107, 111, 115–117
Silicon, 2, 25, 180, 235
Silicon oxide, 65, 74, 112, 113, 115, 116, 118, 121, 124, 179
Silver, 26, 27, 44, 46, 51, 52, 120, 180, 235, 278
Singkep, 79
Sodium, 25–27, 43, 124, 152, 153, 180, 182–221, 235
Sodium chloride, 30, 31, 44, 46
Sodium oxide, 74, 116, 121
Sodium sulphate, 250
Sonic surveying, 20, 22, 79, 81
South Africa, 12, 58, 66, 73, 78, 79
South America, 12, 19, 43, 58, 73, 102, 225, 227, 231, 233, 281, 282, 288
Southwest Africa, 12, 13
Spinel, 153
Sponges, 137
Sponge spicules, 64
Springs, submarine, 146, 147
Stassfurt potash deposits, 31
Stephens Passage, 15
Strata underlying soft sea-floor sediments, 84–102
Streams, 146
Strontium, 25, 51, 124, 152, 180, 182–221, 231, 235, 238–241
Strontium sulphate, 46
Subic Bay, 251
Submarines, 252
Submerged beaches, 11–12
Submerged Lands Act, 283
Submersibles, 89
Suction dredge, 243
Suction head of deep-sea hydraulic dredge, 265–267
Sulphate, 26
Sulphite, 42
Sulphur, 3, 25, 44, 46, 55, 91, 93, 95, 96, 252
Sulphur dioxide, 34
Sulphuric acid, 32, 43
Sumatra, 79, 102
Surf, 9
Surficial sediments, 1

Sweden, 30

Tahiti, 120, 132, 227
Taranake, 19
Television camera, 225, 267
Terrigeneous sediments, 182–221
Territorial sea, 285, 287
Texas, 32, 34, 36, 56, 283
Thailand, 78, 81, 248
Thallium, 26, 181
Thirty Mile Bank, 64, 65, 70
Thorium, 26, 45, 135, 154, 181
Thorium oxide, 47
Tidal currents, 9, 15
Tides, 9
Tin, 25, 46, 51, 78, 79, 81, 120, 124, 126, 181, 231, 238–241, 248
Tin oxide, 47
Titanium, 19, 25, 51, 180, 182–221, 235, 275, 277–279
Titanium oxides, 16
Tourmaline, 123
Travancore, 12
Tribromoaniline, 32
Tricalcium phosphate, 60, 118
Trinidad, 102
Trough, 9
Truman, H. S., 282, 286
Tug-type vessel, 253
Tungsten, 26
Tunicates, 50, 51, 120
Tunisia, 58, 66

Uranium, 25, 44, 45, 61, 181
Uranium oxide, 46
U.S.A., 12, 18, 19, 28, 31, 34–36, 38, 39, 41, 43, 49, 55–67, 69–73, 75, 76, 77, 80, 95, 98, 101, 102, 117, 132, 161, 231–233, 236, 274, 281, 282–284, 286, 289

U.S.S.R., 30, 58, 66

Vanadium, 25, 50–52, 120, 123, 124, 126, 180, 235, 275, 277–279
Vanadium pentoxide, 46
Vein deposits, 96–98
Venezuela, 102
Volcanic debris, 111, 120, 122
Volcanic eruptions, submarine, 146
Volcanic mud, 141
Volcanic rocks, 146, 148, 182–221

Wad, 152
Washington, 16, 72, 102
Water, 65, 74, 115, 116, 121, 179, 182–221, 231, 233, 238, 241
Waves, 8, 10
Weather, 271
Weisbach equation, 265
Well brines, 37
West Virginia, 31
Wind, 9
Wire-line dredges, 243, 245–247
Wolframite, 7, 181
Worms, 144

Xenon, 26
X-ray analysis, 152, 180, 182–221, 224

Yellow mud, 106
Ytterbium, 24, 180, 235
Yttrium, 25, 124, 180, 235

Zeolites, 120, 123, 135
Zinc, 25, 47, 50, 51, 120, 149, 180, 182–221, 235, 238–241, 275, 277, 278
Zircon, 7, 12, 18, 19, 123, 124, 180, 235
Zirconium, 24, 123, 275, 277–279